HISTORY OF SCIENCE SOCIETY PUBLICATIONS

I

THE BLACK DEATH AND MEN OF LEARNING

THE BLACK DEATH
AND
MEN OF LEARNING

BY

ANNA MONTGOMERY CAMPBELL

ASSISTANT PROFESSOR OF HISTORY
NEW JERSEY COLLEGE FOR WOMEN

NEW YORK
COLUMBIA UNIVERSITY PRESS
1931

Published October, 1931

Printed in the United States of America
Kingsport Press, Inc., Kingsport, Tennessee

TO MY FATHER

THEODORICK PRYOR CAMPBELL

AND MY MOTHER

ANNA MONTGOMERY CAMPBELL

TO WHOM I OWE MY LIFELONG ATTACHMENT

TO THE CULTURAL ACHIEVEMENTS OF THE PAST

INTRODUCTORY NOTE

By the generosity of the Carnegie Corporation of New York, a sum of money has been appropriated to the History of Science Society for the purpose of enabling the Society to publish, or assist in publishing, important contributions to knowledge.

The works of scholars and other documents of historical or scientific value are often not sufficiently remunerative to secure acceptance by publishers, who must consider the probable returns on the investment. The support of such an institution as the Carnegie Corporation gives an opportunity to publish works of this kind. A high standard of scholarship and intrinsic merit is maintained by submitting the proposed publications to competent judges. The work of Miss Campbell is the first book to be produced by the aid of the fund at the disposal of the Society, and other important publications will be issued as soon as possible.

G. S. Brett
Chairman, Publications Committee
History of Science Society

PREFACE

THE Black Death of 1347–1350 has figured largely in histories of epidemics; to its history in detail have been devoted some half-dozen special works; and intensive studies have been made of its social and economic consequences, particularly in England. Partly social and partly in the field of intellectual history are the estimates of mortality among the clergy (here, too, chiefly English) and the results upon the church; the treatments of the plague in art; and works dealing with the decided development brought about by it in ideas and public ordinances of sanitation and hygiene. The only important work so far upon the Black Death purely in the field of intellectual history is that in which, during the last quarter of a century, Professor Karl Sudhoff has been a pioneer: the publication of a number of pest tractates evoked by the Black Death and immediately succeeding pestilences, with efforts to identify the authors and obtain information about them. As a result of these scholarly labors, the physicians of the fourteenth and fifteenth centuries are beginning to stand forth as individuals and as a class, in their writings mirroring the medical theories and practice of their age, and occasionally foreshadowing ideas which we regard as characteristic of modern times.

The object of this study is to treat the Black Death only as it affected the intellectual classes, and the fields of learning in which they labored. Medicine and education, especially the universities, are emphasized, since plague tractates of the period and university records furnish the best source material. On the basis of such information as could be obtained, however, an effort has been made to estimate, at least partially, the effects upon other scientists and sciences, natural, mathematical, and social.

It is difficult to express adequately my indebtedness to Professor Lynn Thorndike. Not only did he suggest the idea from which this work developed, but his scholarship, extensive and intensive knowledge of the field of medieval intellectual history, and the generous aid he extends to those setting out on the same path, leave a lasting impress upon students who are so fortunate as to come under his influence. For such gifts of the spirit, too, intellectual stimulus, critical appreciation, and love of my subject, I owe much to Professor Austin Evans. Another inspiring helper, by interest and example, of those who labor for love in the realm of intellectual history is Dr. George Sarton, whom I wish to thank for his interest in my subject and his kindness in allowing me to use, while still in manuscript, the second and third volumes of his important work, the *Introduction to the History of Science*. Dr. and Mrs. Charles Singer gave me valuable assistance, for which I am grateful, by their suggestions and by works of Mrs. Singer's, both in print and in manuscript; and another English scholar and historian, Mr. George Omond, by his thoughtfulness and influence, opened for me in London memorable avenues of approach to authorities in my field. I have been greatly aided by Dr. Arnold C. Klebs, Professor Dino Bigongiari, and Professor G. A. Betz, who read the entire work in galley proof, and considerably enhanced its value by criticism and suggestion; and by Professor Dana Munro, who kindly read portions of it in manuscript, and gave me the benefit of his opinion.

ANNA M. CAMPBELL

COLUMBIA UNIVERSITY
February, 1931

CONTENTS

CHAPTER I

INTRODUCTION

IN THE spring of 1347 a fearful epidemic of bubonic plague broke out in Constantinople, coming from Asia and continuing in the city for about a year. It spread to Cyprus, Malta, Greece, Sardinia, Corsica, appeared in Sicily in October, and before the end of the year had reached Naples, Genoa, Marseilles, and Dalmatia. Early in 1348 it established itself in southern France, Italy, and Spain, maintaining its hold on these countries through most of the year; by June it was in Paris, in July and August penetrated northern and western France, crossed over into England, and devastated Britain and Ireland till the summer and fall of 1349. This was the period, too, in which it extended itself into German lands and was at its worst in the Netherlands. North Germany, Greenland, and Iceland suffered especially in 1350, and Russia from then till 1352.[1]

Estimates of the mortality vary greatly, ranging from one-fourth to nine-tenths of the population; some of them are for a single country, some for Europe as a whole. Among contemporary writers two papal physicians, Guy of Chauliac and Chalin of Vivario, reckon it, respectively, at

[1] The course of the pestilence is worked out in detail by the authors of various works on the subject given in the bibliography, among whom special mention may be made of Philippe, Creighton, Gasquet, Haeser, and Sticker. Excellent tables are given by the last two.

three-fourths and two-thirds; the chroniclers more often approach the higher figure, frequently exceeding it. Recent writers who have attempted to gauge the population loss as accurately as possible also vary widely. Hecker, after careful study of both contemporary and later writers, puts the mortality throughout Europe at one-fourth; Philippe computes it at one-third, and Krafft-Ebing at three-fourths; Seebohm and Gasquet, basing their results largely on records of the English clergy, conclude that one-half of the people of England must have perished.[2]

A masterly description of the nature and course of the disease is given by the distinguished surgeon, Guy of Chauliac, physician of Pope Clement VI in 1348, in his *Great Surgery,* written some fifteen years later. This description is reproduced here, not in exact quotation, since frequent condensation is necessary, but with preservation of the author's phraseology, and even of the person in which he writes.

The great mortality appeared at Avignon in January, 1348, when I was in the service of Pope Clement VI. It was of two kinds. The first lasted two months, with continued fever and spitting of blood, and people died of it in three days. The second was all the rest of the time, also with continuous fever, and with tumors in the external parts, chiefly the armpits and groin;

[2] Guy of Chauliac's *Grande Chirurgie* has been edited by E. Nicaise, Paris, 1890. The portion dealing with the Black Death occupies pp. 166–172 of this edition. No good edition has been published of the treatise of Chalin of Vivario, more correctly Raymond Chalmelli of Vivario, though Dr. Pansier has edited it and now has it in manuscript. His estimate of the comparative mortality of the pestilences of 1348, 1361, 1371, and 1832 is given in Haeser's *Lehrbuch,* 3d ed., Vol. 3, p. 130. Haeser's work, as well as those of the other authorities cited, will be found in the bibliography.

and people died in five days. It was so contagious, especially that accompanied by spitting of blood, that not only by staying together, but even by looking at one another, people caught it, with the result that men died without attendants and were buried without priests. The father did not visit his son, nor the son his father. Charity was dead and hope crushed.

I call it great, because it covered the whole world, or lacked little of doing so. For it began in the East, and thus casting its darts against the world, passed through our region toward the West. It was so great that it left scarcely a fourth part of the people. And I say that it was such that its like has never been heard tell of before; of the pestilences in the past that we read of, none was so great as this. For those covered only one region, this the whole world; those could be treated in some way, this in none.

For this reason it was useless and shameful for the doctors, the more so as they dared not visit the sick, for fear of being infected. And when they did visit them, they did hardly anything for them, and were paid nothing; for all the sick died, except some few at the last who escaped, the buboes being ripened.

Many were in doubt about the cause of this great mortality. In some places, they thought that the Jews had poisoned the world: and so they killed them. In others, that it was the poor deformed: and they drove them out. In others, that it was the nobles: and they feared to go abroad. Finally they reached the point where they kept guards in the cities and villages, and permitted the entry of no one who was not well known. And if powders or unguents were found on anyone the owners, for fear that they were poisons, were forced to swallow them. [Here follow a discussion of what the author considers the true causes, and a résumé of methods of prevention and cure that were employed; since these are treated fully later, they are omitted here.]

And I, to avoid infamy, dared not absent myself, but with continual fear preserved myself as best I could by means of the above-mentioned remedies. Notwithstanding this, toward the end of the mortality I fell into a continuous fever, with a tumor in the groin. I was ill for nearly six weeks, and was in such great danger that all my associates thought that I would die; but the tumor being ripened, and treated as I have said, I escaped by the will of God.

The historical importance of such a catastrophe has received careful consideration from those who have devoted time to its study, and recognition from authors of more general historical and philosophical works. The tendency among these writers is to regard it as epochal in its nature, precipitating processes which might otherwise have worked themselves out naturally, and sharply altering existing institutions and customs. Thus Hecker terms it "one of the most important events which have prepared the way for the present state of Europe," and Gasquet "the real close of the medieval period and the beginning of the modern age." Gasquet is also convinced of its influence upon the religious life of Christendom; Lechner finds it "distinctly a contributory factor in the Reformation"; and Dr. Moll attributes to it serious social disturbances, and a morbid affection of the nervous and intellectual forces of mankind. Another German scholar, Höniger, does not agree with them, considering that the plague was not followed by harmful consequences of great moment, and an English writer, Miss Levett, after studying its effects on the estates of the see of Winchester, comes to the conclusion that the changes it wrought have been exaggerated by economic authorities.

The opinions about the results of this epidemic to be found in more general treatments can hardly be considered here, but two recent works may be mentioned in which it is taken to mark the close of one era and the beginning of another. Hilaire Belloc, in his *History of England*, makes it the dividing point between the medieval and the modern periods; and Egon Friedell, in his *Kulturgeschichte der Neuzeit*, expresses his belief that "the year of the conception of modern man was the year 1348, the year of the Black Death." The most extensive claims for its effects that I have seen are made by A. L. Maycock, who, in *A*

Note on the Black Death, maintains that "the year 1348 marks the nearest approach to a definite break in the continuity of history that has ever occurred."[3]

[3] Hilaire Belloc, *History of England,* London, 1925–1928, Vol. 2, pp. 13, 408; Vol. 3, p. 7.

Egon Friedell, *Kulturgeschichte der Neuzeit. Krisis d. europäischen Seele v.d. Schwarzen Pest b.z. Weltkrieg,* Münich, 1927, Vol. 1, pp. 62–63.

A. L. Maycock, "A Note on the Black Death," in the *Nineteenth Century,* XCVII (London, January–June, 1925), 456–464.

CHAPTER II

PLAGUE TRACTATES OF 1348–1350 AND THEIR AUTHORS

THE Black Death of 1348–1350 and succeeding outbreaks of pestilence in the fourteenth century[1] called forth from physicians in many parts of Europe treatises on the subject,[2] almost all of them written for the use of the public without thought of profit on the part of the authors. The importance of these writings does not, for our purpose, lie in their medical content: there are long, well-organized medical works dating from the fourteenth as from earlier and later centuries,[3] and it would be difficult to dispute

[1] They occurred locally almost every year for the rest of the century, with more severe general attacks taking place in 1360, 1371, 1372, 1382, 1388, 1398. See Georg Sticker, *Abhandlungen aus der Seuchengeschichte und Seuchenlehre,* Vol. 1: Die Pest. Giessen, 1908–1910, pp. 74–81.

[2] *Archiv für Geschichte der Medizin,* XVII (Leipzig, 1925), 264–266. Of the 281 plague tractates listed by Professor Sudhoff, 77 were written before 1400, and 20 or 21 within five years of the outbreak of the Black Death.

[3] Some of the writers and their works from the tenth to the fourteenth centuries are: Rhazes (al-Razi), *c.* 850–924, *Continens* (*Kitab al-hawi*); Haly Abbas ('Alī ibn Abbas), d. 994, *Liber regius* (*Kitab al-Maliki*); Avicenna (Ibn Sina), 980–1037, *Canon* (*Q'anun*); Averroës (Ibn Rushed), 1126–1198, *Colliget* (*Kitab al Kullijat*); School of Salerno, *De aegritudinum curatione* (twelfth century, in Coll. Salern. II, 81–385); Gilbertus Anglicus (early thirteenth century), *Compendium medicinae;* Guglielmo da Saliceto (thirteenth century), *Summa conservationis et curationis; Lanfranchi* (thirteenth century), *Chirurgia magna;* Bernard of Gordon (d. 1318?),

Neuburger's statement that the whole output of pest writings of the period made small original contribution to medical knowledge.[4]

For many reasons, however, they are valuable to students of intellectual history. The first is their very number, and the scattered sources from which they came, in an age when the tedium and cost of reproduction tended to limit literary effort to masters of their art. Hence, they present an opportunity to get the views of the generality of the profession, either in original works or, in the frequent instances of borrowing, by the changes or remarks introduced by the plagiarist. Another is the circumstances in which they were written, many of them under the stress and terror of the worst outbreak of pestilence the world has ever known, and all by men who spoke from actual experience in fighting the disease. Special interest attaches to the mental outlook, the ideas about things in general, which appear throughout the tractates, giving an insight into the philosophy of life of a particular class of learned men in a particular period: and it may be observed that in many cases this is very sensible and free from prejudice or superstition. In others, however, appear interesting evidences of both, in theories of the cause of the pestilence and in methods of treatment. Then there are medical ideas which, if not new, are rare

---

Lilium medicinæ; Henry of Mondeville (d. 1317–1320), *Chirurgia;* John Gaddesden (1280?–1361), *Rosa anglica;* Guy of Chauliac (1300–1368), *Chirurgia magna.*

[4] Max Neuburger, *Geschichte der Medizin,* Vol. II, Pt. I, Stuttgart, 1911, pp. 424–426. He speaks of the development of prophylactic-dietetic writings growing out of the frightful epidemics of pestilence of the fourteenth and succeeding centuries, especially the Black Death, "welcher durch seine entsetzlichen Verheerungen und kulturschädlichen Folgeerscheinungen selbst in der Geschichte der Seuchen einzig dasteht."

enough to arouse attention, and sometimes seem to result directly from the Black Death. And, last, these writings give us additional information, or sometimes all that we possess, about certain fourteenth-century men of learning.

These studies have been made, not as from a vantage point of six centuries of progress, but in a spirit of sympathy and respect. For not only are the twentieth-century mind and twentieth-century science the result of the labors, often obscure or entirely forgotten, of many generations of such thinkers, but, however outward circumstances may vary, human nature remains much the same, and frequently seems to be affected more by heredity than by heritage. A generation which suffered the ravages of the influenza of 1918–1919 can hardly look with surprise upon the helplessness of the medical profession in the face of a far more terrible, deadly, and widespread pestilence; and an age in which Christian Science attracts an increasing number of devout adherents should hesitate to smile upon the medieval blending of religion with the healing art. Nor does it become us, among whom scientists have been seriously studying spiritualism and Freudianism, to view otherwise than with tolerance and not too much superiority such efforts toward a comprehension of extra-mundane forces as found expression in the astrology of our ancestors.

Of the sixteen medical treatises listed here, six were written in 1348, one probably in that year, five in 1349, two in 1349 or 1350, one in 1350, and one refers its origin to the great pestilence. They are given in what seems to be the order of their composition, under the name of the author when possible, and with such facts as are known about author and treatise. A discussion of the subject matter will follow.

1. Master Jacme d'Agramont. *Epistola de Maestre Jacme d. Agramont als honorats e discrets seynors pahers e Conseyll de la ciutat de Leyda, 24 Abril 1348.*

The epistle of Master James to the lords and council of Lérida, written in Spanish and dated 24 April, 1348, bears the earliest date given in any tractate dealing with the Black Death. The manuscript was found recently in the parochial archives of Verdu, in Lérida, by Dr. Arnold C. Klebs, who is now editing it. Nothing further is known about the author.[5]

2. Gentile da Foligno (died 18 June 1348). *Consilia contra pestilentiam.*[6]

"That divine Gentile of Foligno, prince of physicians of

[5] Dr. Klebs communicated his discovery to Professor Sudhoff, and the first part of it, some thirty lines in print, has appeared in the *Archiv für Geschichte der Medizin,* XVII (1925), 120. The dates which occur in the tractates are kept Old Style here, in order not to interfere with the wording; the same is true of place names in quotations from the manuscripts; as *Leyda* for *Lérida.* See also Klebs, A. C., "A Catalan Plague-Tract of April 24, 1548, by Jacme d'Agramont," pp. 229–232 of *Report du 6e Congrès Internationale d'Histoire de la Médecine,* Leyde—Amsterdam, 1927. Published at Anvers, 1929.

[6] Gentile's shorter *consilia* against the pestilence have been published by Sudhoff in *Archiv,* V (1913), 83–86 and 332–337. For his long *consilium,* the one most often cited here in discussing the contents of his tractates, I have used photostats of the MS in the Medicean-Laurentian Library, Florence, Plut. 90, *supra* Cod. 90, fols. 63r–94r, copied in 1428. I have also used portions of an incunabular copy recently acquired by the Library of the Surgeon-General's Office in Washington, and not listed in any bibliography of early printed editions. Seven other early printed copies, in three editions, are listed by A C. Klebs and E. Droz, *Remèdes contre la peste,* Paris, 1925, pp. 22–23, Nos. 50, 51, 52. Large portions of the long *consilium* have been printed in Latin, with French translation, by A. Philippe, in his *Histoire de la Peste Noire,* Paris, 1853.

our and his own time,"[7] was the son of a Bolognese physician, and at first practiced medicine in that city and in Perugia. The annals of Perugia show that in 1325 he was called to lecture on medicine in the university there,[8] and in 1337 he was invited to Padua, lecturing in that university till 1345. Peter Paul Vergerius, who was connected with the Paduan court in the late fourteenth and early fifteenth centuries, and wrote *Lives of the Princes of Carrara,* says that the illness of the ruler of Padua, Ubertino of Carrara, was the reason for summoning to Padua "Gentile of Foligno, illustrious physician of that time.[9]" Ubertino was so impressed by Gentile's elegance and the splendor of his renown that, by his advice, he had twelve of the youths of Padua who showed most aptitude in their studies sent to Paris at his expense, that there they might acquire the best possible training in liberal studies in order later to devote themselves to medicine. All who were chosen, Vergerius adds, completed their training, and became honorable and famous men.[10] The notion of Platina

[7] M. Savonarola, *Commentariolus de laudibus patavinis,* written at Padua in 1440 or within a few years of that date. Michael Savonarola was himself a physician of note, and was grandfather of Girolamo Savonarola. This work was published by L. A. Muratori, *Scriptores rerum Italicarum,* XXIV (1738); the passage quoted occurs in column 1155.

[8] P. D. V. Bini, *Memorie istoriche della Perugina Università degli studj e dei suoi professori,* Perugia, 1816, p. 156.

[9] Petrus Paulus Vergerius, *Vitae principum Carrariensium,* in Muratori, XVI, col. 168.

[10] Two of these youths were probably Enrigetto Lio and Anthony Lio, both of Padua, who became members of the Paduan faculty of medicine about 1350 and 1351 respectively. Enrigetto, in a document of 1351 of the University of Padua, is called "doctor of Paris"; and Savonarola (Mur. XXIV, 1166) says that Anthony had studied liberal arts at Paris. A. Gloria, *Monumenti della Università di Padova,* Padua, 1888, Vol. 1, pp. 448, 371. Another may have been Matthew, son of Master Albertinus de Porta of Padua, who,

that Gentile was physician to Pope John XXII is doubted by Marini, by Tiraboschi,[11] and by Pansier,[12] on the ground of lack of evidence.

Gentile left a number of medical writings which were published in the fifteenth and sixteenth centuries. Some were commentaries on portions of the works of Galen, Avicenna, and Giles of Corbeil, but he is best known through his *consilia,* or medical casebooks, which Neuburger credits with independent thought and observation.[13] His pest *consilia* show both, and also impatience with subtleties and theorizing when practical measures are urgently demanded. From them it is evident that he was active in combating the epidemic, laboring public-spiritedly and unceasingly to assist his stricken fellow-citizens. When the pestilence broke out in Genoa, Gentile addressed his consilium to the college of physicians in Genoa and to certain Pisans.[14] Following this tractate in the Leipzig manuscript in which it occurs is another, signed Gentilis and very like the first, which may have been written for the Neapolitans, as it begins: "It seems evident that the

---

in the roll of the medical faculty of Paris sent to Clement VI in 1349, asked for a canonry in the church at Padua. H. Denifle, and E. Chatelain, *Chartularium Universitatis Parisiensis,* Paris, 1889–1897. Vol. 2, No. 1164.

[11] Girolamo Tiraboschi, *Storia della letteratura italiana,* Milan, 1822–1826, Vol. 5, pp. 387–389.

[12] P. Pansier, "*Les médecins des papes d'Avignon,*" *Janus,* XIV, 405.

[13] Neuburger, *Geschichte der Medizin,* II, 486.

[14] Published by Sudhoff, *Archiv,* V, 332–333. The tractate is taken from a Leipzig manuscript of the *consilia* of Gentile, finished in Bologna in 1488, MS 1178, fols. 53r–54r: "Consilium magistri Gentilis primum in pestilentia quae accidit Januae . . . et direxit istud consilium magister Gentilis collegio medicorum de Janua tempore magnae pestilentiae.

"Doctissimis amicis meis de Janua, ubi prior manifesta fuit haec pestilentia, et quibusdam Pisanis dicimus . . ." As the pestilence cause of the terrible death which first appeared at Genoa, then came to Pisa and Piombino, and which is now at Naples. . ."[15]

To protect Perugia, where he was at the time, Gentile took steps in collaboration with the "venerable college of masters of Perugia." After invoking divine aid, they drew up a set of directions as regarded food and drink, purgation, blood-letting, medicines, and disinfection. The document concludes with the recommendation that the official authority appoint substantial citizens to hold conferences with the physicians and, in accordance with their advice, manage public affairs in so far as the safety of the people was involved.[16]

But nothing availed against the epidemic, and in the stricken city the great Gentile spent himself unreservedly in services to the sick, with the sad result recounted by Francis of Foligno:[17]

---

probably broke out in Genoa toward the end of 1347, this tractate may antedate that of Jacme d'Agramont.

[15] Sudhoff, *Archiv*, V, 333–334.

[16] This tractate is printed by Sudhoff, *Archiv*, V, 83–86, taken from the Würzburg manuscript M.p. MS f. 6, Manual of the Canon at Neumünster, Michael de Leone, fols. 63r–64r. Sudhoff notes that it is almost identical with the third of Gentile's printed *consilia*, and thinks it was probably the last writing from the hand of the great physician. The passages of interest quoted here appear on pp. 84 and 86 of the *Archiv:*

"Divino igitur auxilio invocato praevidit Gentilis de Fulgineo cum venerando collegio magistrorum de Perusia de praeservatione et defensione a tanta pestilentia hoc modo."

"Consulitur autem collegio, quod dominus, cui cura est de ipso, ordinet aliquos bonos homines, qui colloquia habeant cum medicis et secundum eorum informationes disponent civitatem in quantum pertinet ad securitatem hominum."

[17] Francesco di Filippo da Foligno. He is given in the annals of Perugia as lecturing there in medicine in 1351. P. D. V. Bini, *Memorie* (1816), p. 186.

After this in the month of June the revered master drew up prescriptions of these pills which should be taken thrice weekly. . . . And afterward Gentile became ill from too constant attendance on the sick. This was on the twelfth of June. And he lived six days and died, may his soul rest in peace. And this was 1348. And I, Francis of Foligno, was present at his illness, and never left him till his death; and he was buried at Foligno[18] in the place of the Hermits.[19]

3. John of Penna (fl. 1344–1387). *Consilium magistri Johannis della Penna in magna pestilentia . . .*

John of Penna was a member of the faculty of medicine of the University of Naples from 1344 to 1387,[20] composing his tractate there in 1348.[21] It was while the pestilence was raging at Naples, as we have seen above, that Gentile wrote one of the *consilia* just quoted.[22] Thus it is interesting to note that John of Penna opens with the statement that his

[18] Bini (1816), p. 157, says that the "Quadriregio del Frezzi ci assicura del pari del trasporto fatto del cadavere di Gentili da Perugia a Fuligno."

[19] This is printed at the end of one of Gentile's *consilia,* and is quoted by Sudhoff, *Archiv,* V, 87, and by Tiraboschi *Stor d. lett. ital.*, V, 389. Tiraboschi comments on the fact that Alidosi says, without proof, that Gentile died at the age of eighty, and was buried at Bologna. Professor Thorndike found two manuscripts of Gentile's works which stated that they were written in 1359 and 1362 respectively. Lynn Thorndike, *History of Magic and Experimental Science,* New York, 1923, Vol. I, p. 164, note.

[20] G. M. Monti, author of the second section, "L'Età Angioina," of the *Storia della Università di Napoli,* Naples, 1924, p. 85. His name is here given as Giovanni da Penne. He is mentioned in the *Registri Angioina del R. Archivio di Stato di Napoli,* in the years 1344, 1345, 1346, 1349, 1382.

[21] Sudhoff has published two fifteenth-century copies of this tractate. The first, *Archiv,* V, 341–348, is taken from Leipzig Univ. Lib., MS 1178, fols. 54r–57r; the other, *Archiv,* XVI, 162–167, entirely different from the first, comes from Wiesbaden Codex 61, fols. 50r–51r. Professor Lynn Thorndike has found a Joannis de Péine, *De regimine tempore pestis,* in Vatican Palatine Latin MS 1367.

[22] See above, p. 11.

treatise is written in the great pestilence after the great promulgation of the said *consilia,* and especially in view of the said Master Gentile. He opposes the opinion or *consilia* of Gentile, first as to the cause of the pestilence, and then, point by point, as to its prevention and cure.[23]

This controversial document was taken up later, according to another manuscript copy, by Master Francinus of Bologna,[24] through whom it was made public to the University of Paris, with animadversions of an anonymous author against it.[25]

4. Faculty of Medicine of Paris (Oct., 1348). *Compendium de epidimia per Collegium Facultatis Medicorum Parisius.*[26]

[23] *Archiv,* V, 341–342. "Consilium magistri Johannis della Penna in magna pestilentia post magnam promulgationem dictorum consiliorum et specialiter in mente dicti magistri Gentilis.

"Magister Johannes della Penna de Neapoli, famosus doctor tempore magnae pestilentiae, scil. anno domini 1.3.4.8. promulgavit consilium suum quod incipit: Licet praesentis pestilentiae pravitas etc., improbans opinionem sive consilia Gentilis."

[24] A tractate on pestilence by Francinus of Bologna was published by Sudhoff, *Archiv,* VI, 328–333. Written in a fifteenth-century hand, it came from a fourteenth-century manuscript of the Leipzig University Library, MS 1179, fols. 28v–30v. Sudhoff comments that the tone is that of a professor to his pupils.

[25] This manuscript, also an early fifteenth-century copy, was found by Sudhoff in the Breslau Univ. Lib., Cod. III, f. 6. fols. 191–198. Here the author is called Johannis de Pinna, and the date on which Master Francinus of Bologna made this work public to the University of Paris is written 1305, for which Professor Sudhoff suggests 1395. *Archiv,* XVI, 162.

[26] This was first published by J. F. C. Hecker, in his *Wissenschaftlichen Annalen der gesamten Heilkunde,* Berlin, 1834, 10 Jahrgang, Vol. 29, pp. 219–240. He combined a defective seventeenth-century copy in the Bibliothèque Nationale, Anc. fonds latin, MS 7026, with another manuscript of the Bibl. Nat., No. 7082. Later L. A. J. Michon, in *Documents inédits sur la grande peste de 1348,* Paris, 1860, pp. 49–70, published the defective manuscript 7026. Robert Höniger, *Der schwarze Tod in Deutschland,* Berlin, 1882, pp. 152–

This consultation, drawn up at the command of King Philip VI, is the oldest document of the medical faculty of Paris in existence.[27] It is a more pretentious work than any of the others (unless we except that of Ibn Khātimah and the long *consilium* of Gentile of Foligno)[28] longer, and is written from a philosophical standpoint rather than from that of actual experience; in this respect it differs from most of the works produced while the Black Death was raging. The note is struck in the opening words: "Upon seeing effects whose cause is hidden from even the most highly trained intellects, the human mind is driven to wonder, especially as there is in it an innate desire for apprehension of the good and the true; whence everything seeks the good and yearns for knowledge, as appears plainly in the words of the philosopher."[29] This,

---

156, printed the first part, about a fourth of the whole, using two fourteenth-century manuscripts of the Stadtbücherei at Erfurt, Cod. Ampl. Q. 194, fols. 65–67, and Cod. Q. 193, fols. 86–90. Six years afterwards H. E. Rebouis, *Étude historique et critique sur la peste,* Paris, 1888, published the first complete text from a single manuscript, MS 11227, Fonds Latin, Bibl. Nat., giving also a French translation. He refers to the Erfurt MS Q. 194, listed in Dr. Schum's catalogue, without knowing that Höniger had already published about a fourth of it, as Michon had not known of Hecker's work. Professor Sudhoff, *Archiv,* XVII, 65–76, discusses the printed editions, and lists five other Latin manuscripts and a French one copied in the fourteenth century. He includes two French translations of the fifteenth century to which Pansier has called attention, and quotes somewhat at length from some of the manuscripts. Another is mentioned later in the same volume, pp. 257–258. D. W. Singer, *Some plague tractates,* London, 1916, pp. 20–21, discusses a manuscript copy, dated 1373, in the British Museum, Harleian 3050 (XVII), fol. 66 recto(b) to fol. 68 verso(b). See also Chart. Univ. Paris, Vol. 2, No. 1159.

[27] Rébouis, *Étude,* p. 34.

[28] See above, p. 9, and below, pp. 18–21.

[29] "Visis effectibus quorum causa latet etiam ingeniosissimos intellectus, mens humana in admirationem deducitur, et cum ei insit apprehensionis boni verique innata cupido, unde omnia bonum

at a time when, in the terror of the great death, the living had ceased their ordinary ways of life and new burying-grounds were being sought for the dead, shows the power of the philosophic mind. It was the voice of authority, and was heard with respect and repeated with variations throughout western Europe during its own and succeeding generations. The practitioner of Montpellier[30] within less than a year handed back to the University of Paris large portions of its own wisdom. In the anonymous manuscript beginning "First about the epidemic,"[31] appear verbatim many of the astrological explanations; and the poem of Simon of Covino,[32] of 1350, shows marked resemblances to the Parisian *Consultation*. Honest in its intentions, but in execution more damaging to the faculty than plagiarism would have been, is a contemporary version in Italian, *The Opinion of the Medical Faculty of Paris as to the Causes and Treatment of the Plague.*[33] It is much briefer than the real *Consultation*, and gives an explanation of the

---

appetunt et scire desiderant, ut secundum philosophum apparet evidenter." Rébouis, *Étude*, p. 70.

[30] See below, p. 21.

[31] See below, p. 30.

[32] See below, pp. 30–31.

[33] Appendix to "Annales pistorienses, auctore anonymo synchrono," Muratori, *Scrip. rer. ital.*, XI, 527–528. A translation of this, under the impression that it was the genuine Parisian *Consultation*, was published by J. F. C. Hecker, *Der schwarze Tod im 14. Jahrhundert*, Berlin, 1832, two years before he found and edited the correct text. M. E. Littré reproduced in French the first part of it in an article, "Des Grandes épidémies" in the *Revue des deux mondes* of 15 Jan., 1836, and reprinted it without correction in a collection of articles, *Médecine et médecins*, pp. 75–76, published in Paris in 1872. This version of the faculty's pronouncement amply justifies Hecker's observation that the learned faculty was put in an embarrassing position by the king's command, and got out of it as best it could, and Littré's characterization of the pronouncement as being of a bizarre absurdity.

cause of the pestilence which the Parisian doctors would hardly have sponsored; their own, while not in accord with present-day science, at least was in agreement with the best thought of their time. Their pronouncement found its way into the German language also, in a fourteenth-century redaction, *Opinion of the Most Eminent Masters of Paris upon the Mortality of the Glands;*[34] its translation in the fourteenth century into French, Italian, and German is an illustration of the influence of the Black Death in bringing national languages into use in the fields of science and learning in general.

5. Master Albert. *Consilium magistri Alberti ad pestilentiam in 1348.*[35]

Master Albert's treatise is brief, consisting chiefly of prescriptions for prevention and treatment of the plague, and gives no information about the place of composition or, except his name, about the author. In the manuscript it comes after one of Gentile's pest *consilia,* and Sudhoff thinks it originated in Italy.

6. Alfonso of Cordova. *Epistola et regimen Alphontii Cordubensis de pestilentia. 1348?*[36]

[34] *Sinn der höchsten Meister von Paris für die Sterbung der Drüsen,* or, in another manuscript, *Sin der hogisten meyster von Paris vor dy sterbinge der drüsse adir pestilencia, Archiv,* II, 379–383. Sudhoff prints the text of the former from the Breslau Univ. Lib. Cod. MS I, fol. 334. He has found variations of the *Sinn der höchsten Meister* in other MSS copied in the early fifteenth century. See also *Archiv,* VII, 64–65.

[35] Printed by Sudhoff, *Archiv,* VI, 316–317, from a MS of the Vienna Hofbibliothek, Cod. lat. 2317, fol. 34v.

[36] Printed by Sudhoff, *Archiv,* III, 224–226, from two MSS, both, he thinks, copied in the late fourteenth century: Breslau Univ. Lib., MS 111, f. 29, fols. 111v–112r, and Wolfenbüttel Augustanus (18. 18, Aug., Q), fols. 154v–156v. In *Archiv,* XVII, 101, he says he has been

Nothing is known of Alfonso beyond the facts he gives at the beginning of this work, "I, Alfonso of Cordova, master of the liberal arts and of the art of medicine," and at the conclusion, "Written in Montpellier." It seems reasonable to suppose that the regimen was written in 1348, in view of the opening assertion that he had with tremendous zeal sought the cause and state of the pestilences which had their origin and beginning in the current year of the Lord 1348. He divides them into two sorts: the first natural, its causes being astrological, so that that pestilence was naturally in parts of Italy and in regions beyond the sea, in the angle opposite the triangle of the house of Europe.[37] But it should have ceased within a year, and did cease, for the power of the constellation whereby it was diffused was exhausted. But today it was spread abroad through all Christian regions through a cause which was unnatural and was particularly directed against Christians.[38] At first reading these passages seem to point to the period of recrudescence of pestilence in 1350, but when it is remembered that in parts of Italy, Asia Minor, and Africa the pestilence broke out in 1347, it will be seen that "today" may still be reconciled with "the current year of the Lord 1348."

7. Abū Ja'far Ahmad ibn 'Ali ibn Muhammad ibn 'Ali ibn Khātimah.[39] *Morbi in posterum vitandi Descriptio & Remedia. 1349.*

---

able to find no other manuscript except an anonymous one, Münich Latin MS, 259, fols. 179r–180r, which lacks the introduction.

[37] " . . . et illa pestis fuit naturaliter in partibus ultramarinis in angulo opposito trigona mansionis Europae."

[38] The cause of the second, unnatural pestilence is graphically given later, and does credit to Alphonse's power of imagination. See below, pp. 52–53.

[39] For brief treatment of his life and works see M. Casiri, *Bibliotheca arabico-hispana escurialensis,* Madrid, 1770, Vol. II, pp. 334–

From the kingdom of Granada emanated several tractates concerning the pestilence of 1348. In Casiri's catalogue of Arabic manuscripts in the library of the Escorial he gives[40] excerpts from the *Granatensis Encyclica of Ibn al-Khaṭīb*[41] (1313–1374), a list of famous men of Granada, with an autobiography and titles of thirty-seven of the author's own works. Unfortunately the copy which Casiri used, and which he had found in Granada, contained only parts VII–XI, with even VII defective, so that the list begins with the name Mohamad.[42] In the excerpts which he prints, however, appear the names of four men who wrote on the pestilence: Ibn al-Khaṭīb himself, Ibn Khātimah, Alschecuri[43] of Segura, and Abelbanus[44] of

335; C. Brockelmann, *Geschichte der arabischen Litteratur,* Berlin, 1902, Vol. II, p. 259; Taha Dinānah, *Archiv,* XIX (1927), 27–29. The title used here is the Latin one translated by Casiri from the Arabic. When giving Arabic names as used by Casiri and others, I have kept their spelling. Elsewhere the form is that into which Dr. Sarton has very kindly transliterated them for me.

[40] Casiri, II, 71–121, taken from Escur. Cod. 1668 and 1669.

[41] See below, pp. 26–28.

[42] Brockelmann, II, 262–263, says that all fifteen parts exist in Paris, No. 3347(2); a few parts in the British Museum, Suppl. 666, Escur. 1668–9, Cairo V. 128.

[43] Casiri, II, 89. He was born in 1326, and his full name is there given as Mohamad ben Ali ben Abdalla Alakhamita, commonly Alschecuri. He was royal physician and author of many works: among them one on the medical art, *Postulantium munus;* a treatise on experiments, *Major cura;* and a very learned work on the errors of physicians, *Judaeus perdomitus.* His tractate on the pest has not been edited, but is mentioned by Taha Dinānah and by Müller as existing in Escur. MS 1785 (see Note 46 below). I do not find him in Brockelmann.

[44] Casiri, II, 74. He died in 1363, his full name is given as Mohamad ben Mohamad ben Giaphareus Aba Abdalla, and the title of his pest tractate as *Recta intentio.* It is said here that he was quaestor of Almeria and proconsul of Marchena, was upright and learned, and a skilled grammarian and poet, having written a

Almeria. The pest tractates of the first three exist in manuscript,[45] those of Ibn al-Khatīb and Ibn Khātimah have been translated into German and edited.[46]

Most of what we know about Ibn Khātimah is contained in Ibn al-Khatīb's biographical sketch of him, and in his own works. He was born and lived in Almeria, and was, according to Ibn al-Khatīb, one of the most prominent men of Andalusia, a scholar, and an original thinker. He was a theologian, a physician, and a poet, his writings in the last field being evidently considered by Brockelmann the most important, as he classes him in the division of "Poetry and light literature"; a collection of his poems exists in manuscript, Escur. 419. It may be seen from his treatise on the pestilence that during the plague he remained in Almeria attending his fellow-citizens, and most of the measures which he urges to combat the disease are based on his practical experience.

The tractate was written in early February, 1349, when the plague had been raging in Almeria since 1 June, 1348, and was still continuing unabated.[47] The work is in the form of answers to questions that have been asked the writer, and is long, though of the ten questions Dinānah omits the last four as being purely theological. It is more

---

very elegant poem on theology and another on rhetoric. I have found nothing else about him.

[45] According to Casiri, II, 334-335, they are parts 3 and 4 of the Escur. Cod. 1780. The codex numbers have been changed, since in Hartwig Derenbourg, *Les Manuscrits arabes de l'Escurial,* Vol. III (E. Levi-Provençal, ed.) Paris, 1928, the number is MS 1785, fol. 6.

[46] That of Ibn al-Khatīb (Ibnelhatīb, Ibnu'l-Hatīb) by M. J. Müller in *Sitzungsberichten der Königl. bayer. Akad. der Wissenschaften,* Philos-phil. Cl., 6 June, 1863; and that of Ibn Khātimah (ben Hātima, Ebn Khatema) by Taha Dinānah in *Arch. f. Gesch. d. Med.,* XIX, 30–81.

[47] So stated by the author, *Archiv,* XIX, 40.

philosophical than the direct and urgent *consilia* of Gentile of Foligno, except the longest one, and more practical than the compendium of the University of Paris. Geographical and statistical data are given about the disease, as well as he has been able to gather them; moot points are discussed, and though the writer always has his own theory to prove, he adduces facts from nature and experience rather than from written authority. The work is interesting, lucid, and, with one or two exceptions, chiefly where infection is involved, reasonable and intelligent.

8. Physician of Montpellier. *Tractatus de epidemia.* (19 May 1349.)[48]

The *Tractate concerning the Epidemic by a Certain Practitioner of Montpellier*[49] is addressed to the flourishing medical *studium* of Paris and to the whole university of the same place. The introduction declares that, realizing the duty of anyone who knows the cause of that pestilence to use his knowledge for the cure of faithful Christians, the author is going to do what he can for this purpose. The tractate concludes: "Written at Montpellier on St. Ives Day 1349." Nothing more is known of the author, but to judge from his treatise he, if connected with the University of Montpellier, can hardly have been one of its soundest members.

[48] Published by J. F. C. Hecker, *Wiss.Ann.*, XXIX (1834), 240–248; and L. A. J. Michon, *Documents inédits* (1860), pp. 71–81. Rébouis, Études, pp. 35–36, calls attention to the fact that the manuscript from which they both took it, Bibl. Nat. Cod. 7026, is defective, and that a copy of the tractate of the physician of Montpellier also follows the Parisian *Consultation* in the manuscript from which he published the latter, Bibl. Nat. fonds lat., MS 11227.

[49] *Quidam tractatus de epidemia compositus a quodam practico de Montepessulano anno 1349.*

9. *Quaeritur primo quae sint aegritudines nunc currentes* (Early spring 1349).[50]

This treatise is in the form of questions and answers, beginning "The first question is, what are the diseases now current," and ending in the middle of a prescription at the bottom of the page, the next leaf being torn out. There is nothing about the author or the place of composition; the time is indicated by various remarks: that the effect of the conjunction of the planets and other heavenly phenomena will last and is still lasting; that the rains which have been continuous in the whole world since 1346 caused the pestilence to begin in the past summer, which ceased during the heat of the summer, but has broken out again during the dampness of the present winter;[51] that purgatives and bleeding are necessary, "especially in this beginning of spring."[52] The treatise shows thought and first-hand acquaintance with Hippocrates and Galen.

10. John Hake (or Griese) of Göttingen[53] (d. 1349).

That John Hake enjoyed high fame as a physician is attested both by the concluding lines of the pest tractate which contains his name,[54] "These things I have from the bishop of Freising, whom I consider the best physician in

[50] *Archiv,* XI, 51–55.

[51] *Archiv,* XI, 53.

[52] *Archiv,* XI, 54.

[53] For his life, see K. Wenck, "Johann von Göttinger, Arzt, Bischof und Politiker zur Zeit Kaiser Ludwigs des Bayern," *Archiv,* XVII, 141-156.

[54] The text of the treatise is printed by Sudhoff, *Archiv,* V. 37-38. It is taken from Codex Q. 194 of the Erfurt Amploniana, fol. 68r, following the Parisian consultation, and written in the same fourteenth-century hand. Sudhoff gives some facts about John's life, and says that he has come across a whole series of manuscript traces of him, one of them another short treatise on the pestilence in the Breslau Univ. Libr., MS IV, f. 24, fol. 343v, col. 2.

the whole world," and the passage concerning him in the chronicle of the bishops of Freising, that he was "a very fine physician"; here he is also credited with being exceedingly learned in astrology.

He is said to have been the son of a citizen of Göttingen, is known to have attended the medical school in Montpellier in 1314, and Pope John XXII, in a document of 1319, states that he is professor of arts and of medicine. He was probably at that time already physician of Cardinal Jacob Gaëtani, and until the latter's death about 1341 remained under his protection. In 1331 he received the bishopric of Verden, but returned to Avignon, and at the death of Cardinal Gaëtani he attached himself to Cardinal Annibaldo of Ceccano. In 1341 he was made bishop of Freising, but, according to the chronicle of the bishops of Freising, he never went there, but died and was buried at Avignon. His death, occurring shortly before 7 October, 1349, is the only way we have of dating the tractate, as he was still living when his admirer wrote down his directions. The tractate is short, and consists only of prescriptions for medicine, bleeding, disinfection, and so forth.

11. *Treasure of Wisdom and of Art* of five Strasburg physicians (1349).[55]

This work was made known by Ernest Wickersheimer in a paper which he read before the Third International Congress of the History of Medicine in London, 1922, "La Peste Noire à Strasbourg et le régime des cinq médecins strasbourgeois."[56] The names of the authors are given as

[55] Taken from the library of the castle at Berleberg, Codex F4, fol. 209.

[56] It begins: "In dem namen unsers heren Jhesu Christi, so vahet an der schatz der wijsheit und der Kunst verborgentlich." In the choice of title of this work I have followed Professor Sudhoff.

Albertus, Rudolphus, Heinricus von Saisgen, Bernhart von Rostogk, and Henricus von Lubelck, masters of medicine living at Strasburg and humble servants of the city of Strasburg. Their directions are based on much practice and study of the art of all medicine, and they are issuing them because they have observed the need of that city and region, and also because of a desire to protect their lords, the master and council of the city. In the communal archives of Strasburg Wickersheimer found items about four of the five physicians named, and as the first, Albert of Parma, was dead in 1355, and as there was, after 1349, no outbreak of pestilence in Strasburg till 1357, he places the date of composition at the time of the Black Death in Strasburg, July to October, 1349.[57] Albert was canon of Saint-Thomas in 1336, was living in March, 1353, and was dead in September, 1355. The second of the authors Wickersheimer identifies with Rudolf Swinninger, given in the Strasburg archives as "Rudolphus dictus Swenger de Schondorf, physicus, civis Argentinensis," who sold a piece of property in 1349, and in 1380 was dead. Although Professor Sudhoff mentions two Henrys of Saxony listed at this time as physi-

---

Wickersheimer's paper was published in the *Proceedings* of the Congress, Antwerp, 1923, pp. 54–60. For Sudhoff's work on the subject see below, note 57.

[57] This paper did not come to the attention of Professor Sudhoff, and in May, 1924, *Archiv*, XVI, 12–20, he published this work, from the same manuscript Wickersheimer used, together with two pest tractates following it, one of which he considered part of the first. He punctuated the names differently from Wickersheimer, making the first three one, and as, in consequence, he did not have the date of Albert of Parma's death, he was inclined to place the tractate between 1360 and 1370. Wickersheimer defended his position in *Janus*, XXVIII, 369–379, and in *Archiv*, XVI, 187–188, Sudhoff says he thinks the facts that there were five authors and that the date was 1349 are probably correct. He also had given, *Archiv*, XVI, 19–20, data about two Henrys of Saxony, and Bernhard of Rostock.

cians at Strasburg, from Nordhausen and Bernburg respectively, it is the latter only whom Wickersheimer considers as the third physician. He appears in the archives as "Henricus de Saxonia de opido Bernburg Magdeburgensis, physicus Argentinensis," and has already been the subject of a study by Wickersheimer, *Henri de Saxe et le "De secretis mulierum."*[58] Bernard of Rostock appears several times in Strasburg documents between 1356 and 1376; once he is referred to as "Bernhart der Artzot," and once as "Phisicus, prebendarius chori ecclesie Argentinensis." Henry of Lübeck seems to have sunk into oblivion without a trace but for this tractate. It is, according to Wickersheimer, the oldest monument of the medical literature of Strasburg.

12. *Is It from Divine Wrath That the Mortality of These Years Proceeds?* (c. 1350)[59]

This inquiry into the cause of the pestilence was written toward the end of the Black Death, but while it still continued, as is shown by numerous passages in the text: the opening words, "the mortality which has occurred during these years"; a reference to "the present Pope, Clement VI"; the passage "it is evident, moreover, that often the said mortality has lasted in various parts of the world for more than five or six years, and still does not cease"; and others. Germany is indicated as the place of composition by the author's tracing the course of the epidemic in that country from Carinthia to Bohemia; this also shows that

[58] E. Wickersheimer, "Henri de Saxe et le 'De Secretis Mulierum,'" in *Communication faite au 3ᵉ Congrès de l'Histoire de l'Art de Guérir;* London, 17–22 July, 1922, pp. 3–8.

[59] "Utrum mortalitas, que fuit hijs annis, fit ab ultione divina propter iniquitates hominum vel a cursu quodam naturali. . . ." Printed by Sudhoff, *Archiv,* XI, 44–51, from the Erfurt Amploniana, Codex Q. 230.

the time could not have been before 1349. The argument occupies 295 lines in the printed text, and is interesting in that it bases itself largely upon natural science, and that it advances earthquakes as the main cause of pestilence.

13. *Abū 'Abdallāh Muhammad ibn 'Abdallāh ibn Sa'īd ibn al-Khatīb Lisânal-dīn* (1313–1374)[60] *A Very Useful Inquiry into the Horrible Sickness.*[61]

Ibn al-Khatīb was the most prominent statesman and writer of Andalusia in his time. He was born in Loja, Spain, where his father was wezîr, and upon the latter's promotion to political office in Granada, the son studied there. At his father's death in 1341 he became secretary to the wezîr Abū-l-Hasan ibn al-Gaijāb, and when this official died of the Black Death Ibn al-Khatīb succeeded him, thus owing his advancement to the epidemic. According to Brockelmann it was the pestilence which turned him to popular medical writing: his first work had been a description of his journey to Africa in 1347, his second was a manual of medicine dedicated to his sultan. This seems to have been the period in which was written his inquiry into the pestilence, a work composed, he tells us, in the time it took him to dictate it. His bold statement there, that when observation, experiment, and intelligence conflict with the laws of religion based upon tradition the

[60] The fullest information I have been able to obtain about Ibn al-Khatīb is from Brockelmann, *Geschichte der arab. Litt.*, II, 260–263, and his later work of the same name, 2d edition, Leipzig, 1909, pp. 214–216. He is discussed by Casiri, Bibl. Escur. II, 334, and mentioned by Müller (see below, note 61), and Taha Dinānah (above, note 46).

[61] Translated into German and edited by M. J. Müller, in *Sitzungsberichte der Königl. bayer. Akademie der Wissenschaften zu München,* II (München, 1863), 1-28. It is one of the three Arabian pest tractates in MS Escur, 1785, fol. 5, given by Casiri as 1780. See above, note 45.

latter require interpretation, is probably one of the passages in his writings which gave support to his enemies in their later persecutions of him as a heretic.

He remained wezîr till 1360, when General Abū Sa'īd, who had been associated with him in office, headed a revolt against the sultan in his absence, and threw Ibn al-Khatīb into prison. He was soon freed and left Granada, to be recalled in the same year when the sultan regained the city. In 1371 his enemies succeeded in having him exiled from Granada, and instituted against him before the Qādi in that city a religious trial on the basis of his writings. The Meriniden sultan 'Abdal'azîg, at whose court in Tlemcen Ibn al-Khatīb had taken refuge, refused to give him up, but after the sultan's death in the following year the former wezîr was taken and imprisoned at Fez. The sultan of Granada sent his wezîr to conduct the trial, but while preparations were being made for holding the court, an old enemy of Ibn al-Khatīb's incited a mob against him as a heretic. They broke into the prison and killed him in 1374.

Brockelmann classes him as a historian, and lists nineteen of his works, which include history, literature, mystical-philosophical writings, and medical works. Most of the last have disappeared, according to Müller, after having had a wide circulation, and this gives especial value to his treatise on the pestilence. His biographies of famous men of Granada enable us to form some estimate of the mortality of the pestilence among the learned men of that country.[62]

A careful reading of his pest tractate leads to the conclusion that it was written in 1348 near the end of the epidemic in Granada, since the present tense is largely used, and the pestilence referred to is that of 1348. Ibn al-

[62] See above, p. 19.

Khatīb and Ibn Khātimah were friends, as is shown both by the former's high praise of the latter in the *Granatensis Encyclica*, and by letters between them which have been preserved; and it is possible that Ibn al-Khatīb's discussion, in which the predominant motive is a defense of the theory of infection, was in reply to Ibn Khātimah's treatise denying it in obedience to the mandate of religion.[63] The haste in which the Inquiry was composed, "without fitting leisure," indicates a response to a sudden impulse.[64]

14. Dionysius Secundus Colle, physician, of Belluna. *About the Pestilence of 1348.1350. and Pestilential Peripneumonia, and Malignant Likewise.*[65]

[63] See below, pp. 56–59.

[64] J. A. F. Ozanam, *Histoire médicale générale et particulière des maladies épidemiques, contagieuses et épizootiques, qui ont regné en Europe depuis les temps les plus reculés jusqu'à nos jours*, 2d ed., Paris, 1835, p. 76., considers the date to be 1348.

[65] "De pestilentia 1348.1350. et peripneumonia pestilentiali, et maligna simul. Ex libello vetusto Dionisii Secundi Colle a me Titiano Colle filio Leonis Ingegnerii collecto." Printed in *Medicina practica sive nova methodus cognoscendorum pariter et curandorum omnium corpus humanum infestari, et grassari passim solitorum praeter naturam effectuum malignorum, et pestilentium*, auctore D. Joanne Colle Bellonensi, Pisa, 1617, pp. 570–576. This large work is a remarkable illustration of the effect of pestilence on medical thought of the age, for it is dominated by the idea and filled with the word *pestilence*. Volume I (592 pages quarto) is devoted to pestilence and malignant and pestilential fevers; Volume II (304 pages) to other malignant and pestilential diseases. Volume I concludes (pp. 567–592) with hitherto unpublished treatises by earlier members of the Colle family for the treatment of outbreaks of pestilence with which they were contemporary:

Dionysius II Colle, *De pestilentia 1348.1350*, pp. 570–576.

Viventius et Bernardus Colle, *De quibusdam epidemicis*, etc., 1440, pp. 573–583.

George Colle (father of John), *De pestilentia Bellunam vastanti*, 1537, pp. 584–586.

Avantius Colle and George his son, *De pestilentia Zaudum vastanti*, 1513, pp. 587–589.

Facts about the life of Dionysius the Second are few; all that we know of him is gleaned from the *Medicina practica* of his descendant, John Colle, in which is printed this treatise, and from the treatise itself. The first Dionysius, his grandfather, was a knight of Frederick II in 1236, his father was named Manfred, and he himself was a physician of Bellona. His son was Dionysius III, physician, who flourished in 1380, and it was a grandson of the third Dionysius, Titian Colle, elected to the council of Belluna in 1441, who preserved this treatise of his great-grandfather's.

The work, in seven short chapters, was written after the pestilence was past, and is entirely devoted to an account of the disease, including remedies which the author had used against it, and those which he had seen used or of which he had been told. Like Guy of Chauliac, he had himself suffered an attack of the pestilence from which he had almost died, and of which he had cured himself by the use of remedies that he remembered from an outbreak of pestilence in his youth. The remedies he gives which were tried by foolish women and people in general in their panic during the Black Death include the revolting compounds which are frequently associated with medieval medicine: with herbs were mixed the horns and hoofs, stercus and urine, liver and brains of horses, mules, dogs, goats, foxes, and hares. Those ingredients which he gives he has, he says, found safer and easier to get. It is worth noting that this is the only one of the sixteen tractates dating from the Black Death in which prescriptions of this nature figure

---

Bernard Colle, son of George, *De pestilentia Auzurdum vastanti*, 1547, pp. 590-592.

John Colle, besides this work, wrote a *Monumenta sinoptica de peste*, etc., Padua 1631; and himself, according to Jöcher, died of the pestilence.

largely, and that they are given as having been tried by the generality of the people at a time when "everyone tried everything for health," when "no aid was thought of which was not immediately tried," and when "each one was physician for himself, and no dealers in medicines were found." The tractate concludes with no more definite help in dating than "The end of the little book of the pestilence 1348. 1350."

15. *First about the Epidemic* (1349 or 1350).[66]

This treatise was written while the great pestilence continued, probably toward its end, to judge from the assertion that the effects of the conjunction of 1345 "will last as long as it pleases God most high." The astrological introduction is evidently taken from the Parisian *Consultation;* the rest of the work is brief, with emphasis upon diet and prophylaxis.

16. Simon of Covino. *Concerning the Judgment of the Sun at the Banquet of Saturn* (1350).[67]

Simon of Covino, at the conclusion of his poem as published by Littré, is called a scholar of the diocese of Liège; and the manuscript in the Bibliothèque Nationale says that the poem was composed at Paris in 1350. While completed at Paris, it seems to have been begun at Montpellier; the latter is given as the place of composition by Simon of

[66] "Primo de epydimia, secundo de praeservatione eius et corruptione aeris circumstantis videndum est." It was printed by Sudhoff, *Archiv,* V, 41–46, from Erfurt Amploniana Codex Q. 194, fol. 68r. collated with a manuscript of the Bourges City Library, Cod. lat. 184 (167), fols. 138v–139r, in which the date of copying is given as 1360. It exists also in a manuscript of the Königsberg University Library, No. 1958, fol. 100, in a fifteenth-century copy.

[67] *De judicio solis in convivio Saturni.* This poem was edited by E. Littré, *Bibl. de l'École des Chartes* (1840–1841), Sér. I, Vol. 2, pp. 206–243, with introduction by Littré, pp. 201–206.

Phares, court astrologer of Charles VIII, in his *Collection of the Most Famous Astrologers.*[68] Concerning the poem and its author, Simon of Phares says:

> Master Symon de Cuvo, doctor of Paris and great astrologer, who made predictions about the great mortality which was at that time, when it came to pass composed about it a beautiful treatise which he called "de Convivio Solis in domo Saturni"; this is a very singular treatise, and he made it at Montpellier. He predicted also the great and horrible winds which aroused the ocean waves to such an extent that several buildings were overturned by them. And he also foretold, as others had already done, the outcome of the battle between King Philip of France and Edward, king of England, which was so horrible that the sea of Flanders was colored by blood for some days from the multitude of people killed, as is said above.

Simon of Covino practiced medicine and astrology in Montpellier, and Pansier thinks it probable that he had studied and received his doctor's degree at Paris.[69] His poem, as has been said, shows the influence of the Parisian *Consultation,* and is written from the viewpoint of physician and astrologer, the latter more than the former, perhaps because of the demands of poetry. He displays knowledge of what went on at Montpellier when he remarks: "as indeed happened in Montpellier, where there was a greater number of doctors than elsewhere, and nevertheless hardly one of them escaped."[70]

[68] E. Wickersheimer, ed., *Recueil des plus célèbres astrologues et quelques hommes doctes, faict par Symon de Phares du temps de Charles VIIIe.,* Paris, 1929, pp. 220–221.

[69] P. Pansier, "Les maîtres de la faculté de médecine de Montpellier au moyen âge," *Janus,* X, 9.

[70] Here may be mentioned also the poem of Olivier de la Haye (published by Georges Guigue, Lyons, 1888) dealing with the Parisian *Consultation;* and, though composed in 1357, in Dr. Klebs' opinion, and hence a little later than the period to which we are here confining ourselves, the long, versified, French pest regime of

There are various outstanding features about these tractates and their authors which may briefly be summarized. The place of origin was almost exclusively southern Europe: Italy, three; Spain, three; France, five, though of these the author of one was Alfonso of Cordova and of another John of Göttingen; southern Germany, one from Strasburg and another probably from Carinthia; the place of origin of four is unknown. The authors of all but four are given, by name except for the medical faculty of Paris and the practitioner of Montpellier, and are, of course, all physicians.

Two of them, the Parisian *Consultation* and the Strasburg *Treasure of Wisdom*, are the oldest extant medical documents of, respectively, the Faculty of Medicine at Paris and the city of Strasburg. The number of instances of concerted official action against the disease is striking, as that is usually supposed to have begun later in the century. James of Agramont wrote for the council of Lérida; the Faculty of Paris, at the command of King Philip VI; the Strasburg physicians had in mind the council of that city; and Gentile of Foligno and the college of masters of Perugia acted in concert for the protection of Perugia, and made recommendations to the official in charge. The use of vernacular languages for scientific works receives impetus from the Black Death and continues increasingly through the fourteenth century; it is seen here in the Catalan treatise of James of Agramont, the German *Schatz der Wijsheit*, and the Italian and German versions and French translations of the Parisian *Consultation*.

The differences of opinion concerning the cause and

---

the physician John Jacobi, later chancellor of the University of Montpellier. This is published by Klebs and Droz, *Remèdes contre la peste*, Paris, 1925.

nature of the disease helped to introduce something of the scientific spirit in so far as they were supported by the observations and experience of their advocates. Such were the different views of the importance of astrology; John of Penna's attack upon Gentile of Foligno's theory that the particular cause of the disease was the nature of a certain poisonous material generated about the heart and lungs; and Ibn al-Khatīb's insistence upon the forbidden theory of infection, denied by orthodox writers such as Ibn Khātimah.

Chapter III

CONTENTS OF THE TRACTATES

Before taking up a discussion of the contents of the treatises it will be well to have an idea of the attitude of the medical profession of today toward the bubonic plague, as shown in a resumé of modern opinion such as that of Dr. Johns in *Nelson Loose-Leaf Medicine.*[1]

The divisions of the article, much like those of the fourteenth-century physicians, are:

- Historical survey
- Geographic distribution
- Etiology
- Modes of transmission
- Pathology
  - Bubonic type
  - Pneumonic type
- Symptomatology
- Diagnosis
- Treatment
- Prevention

The disease is caused by bacilli found in the parasites of rodents, being endemic in rats and appearing in human beings only after the rat population is decimated. While such an origin of the disease did not become known till late in the nineteenth century and was not suspected by our

[1] *Nelson Loose-Leaf Medicine,* II, 126–130. The article was written in 1920 by Dr. Foster M. Johns.

medieval ancestors, it was noted by Avicenna that a sign of pestilence was when mice and animals living under the earth fled to its surface and were disturbed as if they were drunk;[2] and observations of disturbance in the animal world are made by several fourteenth-century writers.

Modes of transmission, according to Dr. Johns, are infection by inoculation, inhalation, ingestion, or slight mechanical abrasion of the epidermis. Life of the bacilli outside of the living human body or culture media is very limited except in extremely cold climates, so that infection from contaminated articles is relatively unimportant. It is the pneumonic type that is inhaled, but there is rarely a case in which no lung lesions occur. Bubonic plague is ordinarily the forerunner of all outbreaks of the disease in a given locality.

The bacilli pass to the nearest group of lymphatic glands, rapidly multiply, and pass into the blood, whereupon the other lymphatic glands become infected. The heart and kidneys are always the seat of fatty degeneration, and death may occur from heart failure. The action of the bacilli is particularly severe on the walls of the blood vessels, producing frequent hemorrhages in cases of acute blood poisoning. Some poisoning of the blood occurs in almost all cases of plague, certainly in all fatal cases.

The Yersin-Roux anti-plague serum effects a cure in three-fourths of the cases in which it is given within the first twenty-four hours. Medicinal treatment is purely symptomatic, and depressants of all kinds must be used very carefully, with especial attention to keeping up the caloric intake and the carbohydrate balance. Absolute limitation of motion is very essential, particularly for some days after cessation of symptoms.

[2] Avicenna, *Canon,* Lib. IIII, Fen. I, Doc. IIII, Basel, 1556, p. 807.

Most of the section dealing with prevention has to do with protection against rats; personal prophylaxis, however, is strongly advocated in the form of simple bacterial vaccines which should be employed in face of a threatened epidemic.

In considering the medical authorities prior to the fourteenth century to whom physicians at the time of the Black Death turned for assistance, we may take as types the two most influential: Galen[3] (129–199 A.D.), who lived through the great pestilence of the second century; and Avicenna[4] (980–1037), most outstanding of Arabic physicians. They were drawn upon so heavily by the authors of fourteenth-century plague tractates that their influence can better be noted in the course of the discussion of the tractates than by separate consideration.

The subject matter of the treatises is usually divided into considerations of the causes of the pestilence, its prevention, and cure; the last of the three is ordinarily stressed less than the first two for the reason, as several of the writers remark, that the disease almost always has a fatal termination. The causes are customarily treated under the two heads of the general and remote, and of the particular and near; the former being those universal agents responsible for the existence of the disease, the latter the individual factors which affect its entry into the body. Under prevention are grouped suggestions for protection against corruption of the air and infection, and directions for what

[3] The edition of Galen that I have used is that of K. G. Kühn, Leipzig, 1821–1833; it fills Volumes 1–20 of Kühn's *Medicorum graecorum opera quae extant*.

[4] Avicenna, *Liber canonis, de medicinis cordialibus, et Cantica*. The references given here are from the edition of Basel, 1556. In the following quotations, in addition to citing the various divisions, the page numbers of this edition will be given.

Gentile of Foligno calls "the more familiar things and those in which the physician can, as it were, use rule and measure": preservative medicines, food and drink, fasting and repletion, exercise and bath, sleep and waking, and accidents of the soul. The portion devoted to cure embraces treatment of fever and of imposthumes [apostemes].

The stress upon astronomical forces as the real and superior cause of the plague seems to have been largely due to the University of Paris, since they are touched upon lightly or not at all in the tractates which antedated the Parisian *Consultation* or were out of its sphere of influence.[5] Gentile of Foligno, whose death in June 1348 puts all of his writings before that date, pays his compliments to the astrological cause by quoting Avicenna regarding the remote causes, or heavenly figures, impressing themselves upon the near, or earthly figures, in such a way as to render the air putrid and produce pestilence. Astrologers have said, he continues, that this especially happens from lunar eclipses or conjunctions of the planets, particularly Saturn and Mars, under given astrological conditions. The particular causes he mentions as perceptible corruptions existing in the air, but at once strongly states his own position:

> Which of the aforesaid causes it is, however, is of no great moment. It must be believed that whatever may be the case in regard to the aforesaid causes, the immediate and particular cause is a certain poisonous material which is generated about

[5] Galen does not mention astronomical causes. Avicenna gives them briefly and vaguely as the beginning of the changes in the air which bring about pestilence: the first distant cause is heavenly figures, the nearer causes are dispositions of the earth. When the heavenly forces become active and impress themselves upon those of the earth, plague results. *Canon*, p. 806; Lib. IV, Fen. I, Doct. IV, Cap. 1.

the heart and lungs. Its impression is not from the excess in degree of primary qualities, but through properties of poisonousness; whence poisonous vapors having been communicated by means of the air breathed out and in, great extension and transition of this plague takes place, not only from man to man, but from country to country. And, as has been intimated before, it is no great matter in these causes whether it is a constellation or an earthly or aquarian figure, if only we may know how to resist it, and that a stand must be made against it to destroy it lest it destroy us. As for those wishing to extinguish a fire burning a house, it is enough to know that it is a fire, that it may not destroy us, whether it be produced by fire or by motion; and for those wishing to resist the poisonous bite of a dry asp, it is enough to know that the asp was biting, whether it was generated by coition or from putrefaction."[6]

This is almost all he says about the heavenly causes: in the fourth chapter of the same *consilium* there is a brief reply to the seventh of the seventeen doubts there set forth, which concerns the influence of higher bodies—especially the sun and moon—upon the lower. The reply is that such influence exists, resulting from the movements and aspects of celestial bodies; and in answer to the eighth doubt an allusion is made to their effect upon the air.

Astrological factors are not mentioned at all in the brief and practical treatise of Master Albert, and in the careful work of John of Penna only in a few words, "certain conjunctions of the unchanging stars," though we would rather expect a Neapolitan writer to enlarge upon them, especially

[6] Gentile of Foligno, *Consilium contra pestilentiam,* MS in Med.-Laur. Library, Plut. 90 *supra* Cod. 90, fol. 65r–65v. References to treatises of 1348–1350 already located and described, pp. 9–33, will not be given in this section unless more than one copy or edition of the work has been used. Immediately after this, Gentile gives about the same amount of space to a discussion of the air, showing that he does not mean that its corruption is not important, but that it is not of great moment in which of several ways it has occurred. See below, pp. 53–55.

as, in the discussion of symptoms, such phrases as *antecedens causa, conjuncta vera et immediata, causalitas,* have a decidedly astrological ring. Two tractates outside the sphere of Parisian influence were those of Ibn Khātimah and Ibn al-Khatīb. In the long and thoughtful work of Ibn Khātimah the more remote cause is not put first, as in most Christian writings, but follows a full discussion of the nearer cause; and to its astronomical side are devoted barely a dozen lines, so guarded and indefinite as to remind us of modern scientists dealing with a doubtful point. Here, he says, we assume what science asserts, without passing sentence or computing, as is customary in the science of first principles. The first of the three sorts of the more remote causes is the astronomical: change of air through heavenly rays, heavenly lights, and heavenly bodies, undoubtedly by God's decree. Investigation of these has not yet been carried out satisfactorily to men, for acquaintance with astronomical principles, as far as they are of value, is not yet assured, nor even whether they exist. Such is the opinion of upright astronomers. But judgment concerning the inevitable influence of a thing is conditioned upon the truth of its existence. A similar reserve and even greater brevity mark Ibn al-Khatīb's acceptance of the astronomical cause: it depends upon conditions of the heavenly bodies—such as determined conjunctions of the stars—which exercise their influence on the world, as astrologers maintain and physicians assume upon their authority. The astronomical cause, he adds later, is foreign to the object of medicine.

The University of Paris, however, leader at the time in astronomical science, according to Duhem, and stronghold for Christendom of orthodox theology, boldly faced the problem, setting forth a confident explanation of how the conjunction of 1345 brought pestilence upon the earth. On 20 March, 1345, at one o'clock in the afternoon, occurred an important conjunction of three higher planets in the sign of Aquarius, which, with other conjunctions and eclipses, is the cause of the pernicious corruption of the surrounding air, as well as a sign of mortality, famine, and other catastrophes not connected with the present subject. Aristotle is quoted and other, unnamed, ancient philosophers referred to for the fact that the conjunction of Saturn and Jupiter brings about the death of peoples and the depopulation of kingdoms, great accidents occurring on account of the change of the two stars themselves; and Albert of Cologne is adduced in support of the assertion that the conjunction of Mars and Jupiter causes great pestilence in the air, especially when it takes place in a warm and humid sign, as occurred in this instance. For, the Paris doctors explain, Jupiter, a warm and humid planet, drew up evil vapors from earth and water, and Mars, being excessively hot and dry, set fire to those vapors. Whence there were in the air flashes of lightning, lights, pestilential vapors, and fires, especially since Mars, a malevolent planet generating choler and wars, was from the sixth of October, 1347, to the end of May of the present year in the lion together with the head of the dragon. Not only did all of them, as they are warm, attract many vapors, but Mars, being on the wane, was very active in this respect, and also, turning toward Jupiter its evil aspect, engendered a disposition or quality hostile to human life. From this were generated strong winds, which, according to Albert, Jupiter has the property of raising, particularly from the south, giving rise in lower countries to very great heat and dampness; in regions about Paris the dampness was greater than the heat.

This explanation of the higher cause of the plague was taken up by writers all over Europe, and for over a century is to be found repeated in more or less detail in most pest tractates which are not merely sets of medical directions. In the remaining ten, presumably written during the Black Death and after October, 1348, only three omit it entirely;[7] in some, portions are repeated verbatim, in others altered or expanded.

The Italian version in the *Annales Pistorienses*,[8] which purports to be a translation of the text of the Parisian *Consultation,* gives the following rendition: In India and about the Great Sea, constellations, combating the rays of the sun, struggled violently with the waters of that sea, which rose in vapor and fell again for twenty-eight days. At last the greater part of them were drawn up as vapors, those which were left being so corrupted that the fish within them died. The corrupted vapors which had been drawn out could not be consumed by the sun, nor could they be converted into wholesome water like hail and dew, but spread abroad through the air. This had happened in Arabia, India, Crete, Macedonia, Hungary, Albania, and Sicily, and if it should reach Sardinia no one would survive; the danger would continue in countries to which this air had access as long as the sun was in the sign of the lion. But the constellations were by their divine might striving to aid the human race, and, together with the sun, to break through the mists; so that within the next ten days the mists would be changed into noisome rain, purifying the air. From this rain people should protect themselves, and before and after it build large fires in open public places and in their houses.

[7] *Treasure of Wisdom;* John Hake; and Dionysius Colle.
[8] See above, p. 16.

Seven months later the practitioner of Montpellier addressed to the University of Paris a tractate concerning the pestilence,[9] repeating in it lengthy passages from their own *Consultation* in regard to the astronomical cause, but with discursive additions and comments that put it in the class with the Italian version rather than with the Parisian model. For instance, the reason why the epidemic rages in two distant towns and not in places between them is on account of the aspects and rays of the planets which strike these places, like the glance of the eyes upon an object; as, when Saturn looks upon Mars with malignant aspect, or Mars with malignant aspect upon humane Jupiter, then the rays of those planets kill where they strike. Astronomical causes are assigned for other peculiar conditions and manifestations of the plague, as we shall see below; eighty-one of the three hundred lines in Michon's edition of the treatise deal with astrology, most of it of a type analogous to the extract given.

The tractate *First about the Epidemic*[10] gives almost exactly the University of Paris's explanation of the astronomical cause, wisely refraining from enlarging upon it. While the poetic form and classical allusions of Simon of Covino's *Judgment of the Sun at the Banquet of Saturn*[11] class it as much among literary productions as among medical, it is scarcely more a work of the imagination than is the treatise of the practitioner of Montpellier, and is throughout obviously written by a physician of learning and powers of observation. The opening of the prologue gives a description of the conjunction of 1345 and its consequences very similar to that of the University of Paris,

[9] See above, p. 21.
[10] See above, p. 30.
[11] See above, pp. 30–31.

even including the reference to Aristotle's book of the properties of the elements.

Two of the tractates, written perhaps within a year of the Paris *Consultation,* devote only a line or two to the astronomical forces, which they mention as the first or remote cause of the disease, passing on to a fuller discussion of other causes. According to one, that of Alfonso of Cordova,[12] the first cause of the pestilence is a natural one, namely, the eclipse of the moon immediately preceding in the sign of the lion, together with a powerful commixture of unlucky planets. In the other, *The First Question Is, What Are the Diseases Now Current,*[13] we find concerning the remote cause only the terse statement that it is the conjunction of the three higher planets, and several constellations and eclipses, which have occurred, and the effect of which will last, and is lasting till now. One writer only has the hardihood to dispute the astronomical influences. The author of *Is It from Divine Wrath That the Mortality of These Years Proceeds*[14] opposes the opinion that Saturn's being in his own house is responsible for the plague, for he tarries there for three years once every thirty years, but pestilence does not happen every thirty years. Nor does it seem credible that it proceeds from the conjunction of Mars and Saturn, for no conjunction could last more than two years, and the disease has lasted five or six. Some people might argue that the effects could outlast the cause, as the chicken can be hatched when the cock who fathered it is absent or dead. But many conjunctions have occurred without this result, and moreover there is a certain order in heavenly bodies which is entirely lacking in

12 See above, pp. 17–18.
13 See above, p. 22.
14 See above, p. 25.

the progress of the plague, which skips unaccountably north, south, east, or west; wherefore it does not appear to the writer that the stars are its immediate cause. He does not, however, go so far as to deny them mediate or remote responsibility.

Besides the celestial causes, other causes and portents are given, usually regarded as inferior to the first and probably dependent upon them. Earthquakes of 1347 and earlier are mentioned in a few of the tractates with varying degrees of emphasis.[15] The author of *Is It from Divine Wrath* is inclined to the theory that the epidemic was the result of corruption of the air following the earthquake of 1347. This view he supports, since, he says, it is a sure truth that an earthquake is caused by the earth's exhalation of fumes shut up in its inner parts. When these beat violently against the sides of the earth, and there is no opening, they shake and move the earth, as is evident from natural philosophy. So after the earthquake of 1347 the corrupt air escaping infected the air upon the earth and killed people in various parts of the world. Of the eight proofs he gives in support of this hypothesis only the first seems to bear directly upon the earthquake: that, as far as Germany is concerned, the mortality first began in Carinthia after the earthquake there, when the vapor and air escaping from the inner parts of the mountain followed the subalpine valleys, destroying whole towns in the mountains, then going successively to Austria, Hungary, Bavaria, Moravia, Bohemia, the province of the Rhine, Swabia, and other provinces of Germany, following no regular course,

[15] Neither Galen nor Avicenna include earthquake among the causes of pestilence, though the latter verges upon it when he says that when putrefactions take place in the inner parts of the earth, for reasons which we do not know in detail, they reach the water and the air. *Canon*, p. 806; Lib. IV, Fen. I, Tract. IV, Cap. 1.

but going as the wind blew the corrupt air. He sketches and answers six doubts which may be advanced. (1) The pestilence arose in some regions before the earthquake. (2) The vapor would be so scattered and attenuated that people would not be harmed. (3) Pestilences do not follow all earthquakes. (4) Not everyone dies in infected spots. (5) Beasts do not fall victim. (6) The pestilence is less severe in summer than in winter in some places, while in others the opposite is true, and during the dog days "there has been doleful calamity and frequent lamentation of the peoples, and in the same places in the season when the sun's rays were oblique peace and tranquillity prevailed."[16] His six replies are: (1) Such an earthquake must not be expected as that which did so much damage in Carinthia and other parts of Germany, and also those which occurred in the ultramontane regions, Italy and Ravenna. Before the Carinthian earthquake occurred the gases, held captive beforehand for a long time within the closed places of the earth, were compressed within the sides of the earth with violent strainings, since it was not easy for them to lift so great a mass of weight, and through the straitness of the pressure no little gas or vapor escaped beforehand through the surface of the earth in some places, and poisoned the atmosphere there.[17] (2) That vapor is not scanty, and it takes it a long time to vanish and lose strength. (3) It is not necessary for the mortality to follow every earthquake;

[16] His phraseology is at times graphic and poetic. The exact wording of this passage is, "fuit casus flebilis et creber luctus populorum et in eisdem locis tempore obliquitatis solaris pax et tranquilitas habebatur."

[17] ". . . spiritus infra clausuras terre capti prius tempore longo se mutuo infra terre latera artaverunt pressuris fortissimis, quia non erat eis facile tantam molem gravedinis elevare per quam pressure angustiam non parvus spiritus seu vapor per rimas terre prius exalavit in aliquibus locis et aera infecit eorundem."

"the reason," he says, "is sufficiently obvious"; though it does not seem any more obvious than why the pestilence could not be the result of Saturn's tarrying in his own house, because he tarries there every thirty years without producing such dire consequence—an argument which the author has recently advanced against the heavenly cause. (4) People differ in constitution and mode of life, and some offer better resistance than others. (5) Man is nobler and feebler than other animals, and so can suffer what they escape. (6) The poisonous matter is driven to different places at different seasons, after the manner of propulsions of strong poisons.

There is no other such extreme advocate of the earthquake theory, and only in two other works is it mentioned: Alfonso of Cordova makes "that very great earthquake" the second natural cause; and the University of Paris brings it in at the conclusion of a discussion of the corruption of the air, giving as its explanation of earthquakes that "Corruptions of this kind perhaps have been or may be due among other causes to decaying matter piled up in the interior of the earth which sometimes causes earthquake, as in fact it recently has done."

However, while not expending much time upon earthquakes, the University of Paris does go rather fully into the many portents and signs which not only foretell, but by their nature produce, pestilential disease.[18] Change of season is given first place, and the seasons have of late been very irregular, the temperature changeable, and the

[18] Avicenna gives as signs of pestilence: fiery apparitions, such as falling stars, in the autumn; east winds in December, and the air turbid as if filled with dust, rain threatening but not falling, and excessive heat by day and cold by night. *Canon*, p. 806; Lib. IV, Fen. I, Tract. IV, Cap. 3.

whole year, on an average, warm and damp.[19] Alarming phenomena have also appeared: exhalations and flashes such as the dragon and the flying stars; the sky yellowish and the air reddish, on account of the burning fumes; flashes of lightning and blazing lights; great thunderclaps and strong violent winds from the south, disposing bodies more to rapid putrefaction; a multitude of fish, beasts, and other dead creatures on the shore of the sea and in many regions; trees covered with dust; and, according to the observation of some, a multitude of frogs and reptiles generated from putrefaction.[20] All these things seem to precede great corruption in air and earth; moreover, many wise men, held in honorable remembrance, making sure investigation by experiments, have foretold all of them.

Simon of Covino, who, as we have noticed, wrote in Paris and shows the influence of the *Consultation* there, notes that the coming of the plague was marked by natural phenomena, heavy mist and clouds, lightning and falling stars, and deadly south winds in hot blasts. The tractate, *The First Question Is, What Are the Diseases Now Current,* comments on the weather, continuous rains, unseasonable temperature, south winds during the winter and north winds in the spring. In Almeria, Ibn Khātimah attributes the pestilential air partly to irregularity of seasons, in the matter both of temperature and of rains and winds. In the vegetable and animal world are extraordinary occurrences. Plants exposed to the corrupt air become sickly and wither, their leaves looking as if they were covered with dust, and

[19] Galen says that in pestilence a change of air takes place to such an extent that the seasons of the year do not preserve their due order. Kühn, XIX, 392.

[20] Avicenna notes that it is a sign of pestilence when frogs multiply and reptiles are generated from corruption. *Canon,* p. 807; Lib. IV, Fen. I, Tract. IV, Cap. 3.

the fruits rotting before they are ripe. A trustworthy man had told him that Christian merchants coming from the East had seen in the sea opposite the plague-stricken Turkish coasts mangled and rotting fish driven over on the water and piled up in great numbers, spreading evil odors. They thought that these fish had been struck by lightning, but that, declares Ibn Khātimah, is entirely false, as lightning is extinguished by water, and, even if it were not, many bolts would not destroy a great number of fish. It seems to him that water, like air, becomes so changed and defiled that animals within it die.

Of the particular and near causes of the pestilence, the most important is, practically without exception, granted to be corruption of the air; the corruption is explained in various ways, according to the intellectual predilections of the authority speaking.[21] Its most earnest proponent is the Moor, Ibn Khātimah, who devotes a large proportion of his work to this cause, making it all-embracing and others accessory to it. In doing so, he accomplishes the strange feat of denying infection "in which the Arabs in their ignorance (before Islam) believed," but which is now forbidden by the tradition of their religion, while admitting that the disease follows upon association with the sick, use of clothes or bedding that they have had, and admission

[21] Galen's definition of pestilence is a disease arising from corruption of the air, and attacking all or a great many people, resulting in the death of many. XIX, 391. Even more emphatic is his statement that no one with any claim to intelligence can be at all ignorant of the fact that pestilential condition of the air brings on fever. VII, 279.

Avicenna also considers that pestilential fever arises from damp and turbid air, which does not become corrupt of itself, but by the admixture of bad fumes from marshy, stagnant places, or from many unburied corpses, or from the earth. *Canon*, p. 129, Lib. I, Fen. III, Doct. V, Cap. 1; p. 806, Lib. IV, Fen. I, Tract. IV, Cap. 1.

into a town of people from a pestilential place; and further that exactly the same type of plague ensues in the victim as that which afflicted the person from whom it was contracted.

According to his thesis, the corruption of the air may be partial or total. The partial occurs through the deterioration of some or all of its accidental peculiarities, without entire alteration or destruction of the element, by increase or decrease of the number of its peculiar properties, or by mixing and combination with foreign substances. Here the air is not changed in its kind, but only in its nonessential peculiarities; it still contains all of its elementary constituent parts, so that, though corrupted, it cannot be defined differently, and is still air. In the case of total change, however, the elementary constituent parts are completely corrupted by putrefaction, so that the nature of the element is entirely altered, and the definition which formerly fitted it no longer does so, the same name being still given it only in an allegorical sense. It then resembles foul vapors more than it does pure air, not to mention good air. This may be seen in old store-houses of provisions, in the altered black tinge in ships, in wells in which animals have met death and where their bad odor remains shut up. The air in such places is befouled and entirely changed, especially into a fume deadly to animals, and if a man finds himself in it he dies at once. In such a reek no light burns, but goes out at once when lit and put in it, which is the highest degree of foulness and deadliness of air. By this sort of change, or something like it, is the air altered in accounts of epidemics in which in a few days, according to historians, thousands of people died.

The learned Arab knows, however, that he may be

inviting attack, and meets the anticipated charge that what he says contradicts the dictum of science that an element cannot change (since thereby it would become a compound) with the rebuttal that the air which lies next us and in which we live is not a single pure element, but a combination of watery vapor, dry fume arising out of the earth, fine fiery particles, and, chiefly, air. All these are mingled in what we call air. Hence a process of decay can effect change in these materials while it could not do so in air as a pure element. Though if this element were generally in a pure condition, perhaps it might be possible in the upper margin where the rays are reflected.

The causes of this change of air may be of three sorts. (1) The astronomical cause, which has not been satisfactorily investigated as regards its value or even its existence. (2) Change through irregularity of seasons, either in temperature or in rains and winds; this Ibn Khātimah considers the cause of the present epidemic. (3) Change of air as a result of putrid fumes of decaying matter, arising from manure, decaying stagnant bodies of water, low swampy fields with unstirred air, rotting plants and vegetables, corpses left unburied on the battlefield, and plague-stricken cattle.[22] The reason that he attributes the

[22] Galen considers the origin of the trouble to be a great number of corpses left unburied, as happens in war, or the exhalation of certain swamps or ponds in the summer, and unusually great heat. VII, 290.

Avicenna's remarks on this subject seem peculiarly uninspired: The corruption of the air often comes from the earth, in which case one should stay in his room and seek high shelters shielded from the wind. Often the beginning of the corruption is in the air itself, received from the neighboring air, or on account of the heavenly cause, whose quality is not known to men, whereupon refuge should be sought in subterranean dwellings or those surrounded on all sides by walls, and in caverns. *Canon,* p. 129; Lib. I, Fen. III, Doct. V, Cap. 1.

pestilence to irregularity of seasons is because it has spread to most if not all lands, and through all seasons of the year, a statement which he proceeds to support by telling what he personally knows of the disease in Almeria and what has been told him about its origin and course in other places.[23]

[23] His account of the progress and course of the pestilence, while not properly a part of this discussion, seems interesting enough to be included here in a note: it broke out in Almeria 1 June, 1348, raged for part of the spring, the whole summer, and autumn, and into the winter till the present date of writing, the first part of February. [He gives the Arabic dates usually, sometimes with the Christian month, and sometimes only the Christian date.] It continues unabated, and the kind of sickness has remained the same, though the symptoms seem to fit themselves to the seasons. It began lightly, and continued so till the end of September, when it became more violent, as it is still. It broke out in a corner of the city where dwelt the poor and needy, in a family by the name of Beni Danna, spread gradually among their neighbors till it reached the outer borders of the city, then penetrated within. The greatest number of deaths per day is 70, which is rather small compared with that of other cities of Islam and Christendom. Trustworthy witnesses told us that in one day in Tunis 1,202 deaths occurred, in Tilimsan over 700, the other day in Valencia 1,500, on the island of Mallorca [Majorca], on May the twenty-fourth, 1,252; there the survivors were estimated at about a quarter of the inhabitants. The like is reported to us of all larger and smaller cities.

Opinions differ concerning its origin. On the authority of many Christian merchants who came to Almeria it arose in "Hata," which is called China in Persian speech, as I have learned from an authority from Samarkand. China is the limit of the inhabited earth to the east. It spread through China, then to Irak and the Turkish lands. Others say that according to reports of Christian travelers it began in Abyssinia, and from there spread to the neighboring lands and to Egypt and Syria.

These different reports show that the catastrophe has visited universally all lands and zones. The reason for the difference of the reports is that when it appears in lands lying at the edge of the world, its inhabitants think that the disease originated there; and from there this view spreads. It is also reported to us from many sides that it was in the Genoese stronghold Caffa, which was re-

Ibn Khātimah's friend, Ibn al-Khatīb, denies that change of air occurs except as it becomes a vehicle of infection. After mentioning the more remote astrological cause, which he does not consider the business of the physician, he gives as the nearer cause of the pestilence the corruption of the special atmosphere at the place of its entrance, whether issuing from the infected house or through transmission. But "when we speak of the corruption of the special air, we mean, for instance, the air of the house which is attacked, and then the air of the city while the plague is established there, while the air of the neighboring territory remains unaltered; and we exclude the corruption of the air in general, since that next it remains sound. Did a general corruption of the ocean of air take place, as is the case with the water in ponds in which crafty means are employed against the fish, then no air would be left unaltered.

The Christian Spanish physician Alfonso of Cordova, attributes the first year of the pestilence to astronomical forces; but as to why it did not cease then, which he is convinced it should naturally have done, he has dark suspicions that speak more strongly for his imagination and

---

cently besieged by an army of Mohammedan Turks and Romans, then in Pera, then in great Constantinople, on the islands of Armenia to the coasts of the Mediterranean, in Genoa, in France. It spread through Andalusia, the regions of Aragon, Barcelona, Valencia, etc., in the greatest part of the kingdom of Castile to Seville in the farthest west, reached the islands of the Mediterranean, Sicily, Sardinia, Majorca, Iviza, sprang over to the neighboring coasts of Africa, and went from there farther to the west.

This is an example of the wonderful deeds and power of God, because never before has a catastrophe of such extent and duration occurred. No satisfactory reports have been given about it, because the disease is new, and because of the great distances involved, and the absence of travelers of either religion. God only knows when it will leave the earth. It is reported that it has ceased in Majorca and many Mohammedan and Christian cities.

Christian prejudices than for his medical learning. Experiment has shown that this prolongation of the pestilence does not proceed from any constellation and so from no natural infection of the elements, but from depths of malice invented through the subtlest artifice of profound iniquity, threatening especially Christians. Above all, the faithful should be careful about food and drink, and any of the necessities of life which could be infected or poisoned, and should take all sorts of medicine against poison. Pills against infected air are of great value, "since air can also be infected artificially, as when a certain confection is prepared in a glass flask, and when it is well fermented, the person who wishes to do that evil waits till there is a strong slow wind from some region of the world, then goes against the wind, and puts his flask against the rocks opposite the city or town which he wishes to infect, and making a wide detour by going back against the wind lest the vapor infect him, pulls his flask violently over the rocks. When it breaks, the vapor pours out and is dispersed in the air, and whoever it touches will die as if from pestilential air, and more quickly." The pills which Alfonso prescribes against this alarming danger are to be made of the best theriac major and bones from the heart of a stag.

Gentile of Foligno, near the beginning of the first chapter of his longest *consilium*, alludes to the particular and manifest causes of the pestilence as perceptible corruptions existing in the air from near at hand or brought from distant parts, especially from the south: these occur in the opening of wells and caverns shut up too long; or as a result of the shutting up of air in walls or roofs; or they rise from lakes or ponds; or on account of the stercus of animals, corpses, or other evil-smelling putrefactions. In the fourth chapter of the same *consilium*, he counters seventeen doubts which have been raised concerning the epidemic; numbers 3, 4, 6, 8, 9, 10, 17 involve the corruption of the air. Though a simple body, the air can be corrupted because of other substances mixed with it; the pestilence tends to be worse in early summer and late autumn than at other seasons, in spite of the fact that the air is then cool and dry, because the air still retains much warmth, vapors are plentiful, and the pores of the body expanded; bodies with wide pores are more subject to pestilence than those with narrow ones, because the latter become clogged less easily and admit less infected air. He advances two explanations of how the infected air is drawn to the heart. One is that the great amount of poison, breathed in with the pestilential air, multiplies itself while feeding upon the moist humors of the body, at the same time becoming more malignantly poisonous, till it reaches the heart and drives out the vital spirit. The other is that bodies with wide pores admit the pest-laden air through the pores, whence it is drawn to the heart.

There are questions about the peculiar behavior of birds during pestilence; why it attacks people but not cattle and other beasts; and why some people and not others. The birds' erratic movements are due to the air being more subtle than the earth, and the upper air than the lower, so that it is more easily infected, especially when the disease is caused by heavenly forms and images impressed upon the air; and then birds that ordinarily fly high come near the earth. But if the pestilence arises from the earth, as in the case of earthquakes, the infection is worse there, and birds that usually stay near the ground mount aloft. It is the singular property of the air which causes it to impress itself on humans and not on cattle; this may also be caused by heavenly forms and images. Differences of constitution and disposition are largely responsible for difference of susceptibility among people. One further doubt is concerned with the observation that monks and persons who are incarcerated do not die so generally as those who are not shut up. Gentile acknowledges this, but calls attention to the fact that when pestilence does break out among them they are nearly all swept away. The explanation of these two truths does not connect itself in his mind with the presence or absence of rodents and their parasites, but is to be attributed to all the members' having the same regimen, breathing the same air, and living in a place of the same aspect.

The University of Paris, like Ibn Khātimah, believes that the present epidemic proceeds immediately from air not merely altered in its qualities, but corrupted in its substance by noxious vapors mixed in with it. These vapors are chiefly the result of the conjunctions of the planets, but have been made worse by the south winds which have been blowing, bringing, perhaps, vapors from swamps, lakes, deep valleys, and unburied corpses; they may also be partly due to decaying masses piled up inside the earth, such as caused the recent earthquake. This earthquake, as has been seen, is thought by the author of *Is It from Divine Wrath* to have been the immediate cause of the corrupt and poisonous exhalation from the earth, which "has infected the air and suffocated with sure, sudden extinction" people who breathed it.

The author of another set of medical questions and answers, *The First Question Is, What Are the Diseases Now Current*, divides the extrinsic causes into the remote or astrological, and the near, of which the first is the corruption of the air in its substance, and its change in humidity on account of continuous rains. The principal indication of the opinions of the practitioner of Montpellier and of the author of *First about the Epidemic* as to the part played by the air in the spread of the disease is shown by the fact that each begins his disquisition with the explanation of the derivation of *epidemia* as being *epi,* above, and *demos,* containing, or that which comes from above and exists in the surrounding air: each adds, the corruption of the air. Simon of Covino, who, like these two writers, strongly shows the influence of the Parisian *Consultation,* makes the plague follow upon the adulteration and poisoning of the lower air, which has been compressed by the dense, heavy clouds or rain caused by Saturn in the conjunction of the planets. The Strasburg *Treasure of Wisdom* also stresses the danger from the poisoned air.

The third extrinsic cause of the pestilence, the first being natural phenomena and the second corruption of the air, is infection.[24] The clearest statement of the infectious nature of the disease is given, strangely enough, by Ibn Khātimah, who feels himself obliged by the law of his religion to deny the existence of infection.[25] That the evil spreads, he says,

[24] Galen cautions that it is dangerous to have dealings with those who suffer from pestilence, the danger lying in contagion, as in the case of scabies or ophthalmia. It is, besides, dangerous to remain with those suffering from pestilence, and, on the whole, with any who breathe out putrid matter. VII, 279.

[25] Before the time of Muhammad, the Arabs believed that epidemics were the work of divine or demonic agencies. Muhammad hesitated between this belief, a logical consequence of the dogma of divine omniscience and omnipotence, and the practical view based on experience. At first he forbade the possessor of healthy herds to approach those who were sick; then came the command not to go into a land afflicted by pestilence, but if already there not to leave it. However, soon after his death, his successors began to enunciate the principle that pestilence was a punishment sent by God to whom he would, and that it did not come from men. The last naturally precluded infection, and the saying became accepted "Pestilence is a test of faith for every Moslem." From an early

is evident from observation and experience, it having not yet happened that a well man remained long with a sick one without being attacked by the disease. Almost as harmful as the air breathed out by the sick, if not entirely so, are the fumes from their bodies, pieces of clothing, beds and linen on which the sick lay, if they are used again. The author has observed that the inhabitants of a portion of Almeria, where the clothing and bed linen of the sick were sold, died almost without exception, while dealers in other markets under the same conditions fared as other people. He has also observed that cities which forbade the entry of people from stricken places remained spared as long as possible; most of the smitten, living in strongholds about Almeria, date the entrance of the disease from the arrival of someone from a pestilential place, and his death among them.

> The best which long experience has taught me is that when anyone comes into contact with a sick man, forthwith the same disease seizes him, with the same symptoms; if the first sick man has spit blood, he also spits it; if the former was hoarse, he likewise becomes hoarse; if with the first buboes appeared on the groin, so do like buboes appear in the same place with the other; if with the first a boil arose, there arises with the second a like boil; also the second sick man carries the sickness on farther. The family suffers in the same way, shows the same symptoms; if the sickness of one member runs a fatal course, the others suffer the same fate; if the sick man is saved, so they also escape death. In this way on the whole, with slight differences, has it gone in our city.

Yet in spite of this lucid exposition he denies infection, declaring that the spread of the disease from sick to well

---

date leprosy was an exception. E. Seidel, "Die Lehre von der Kontagion bei den Arabern," *Archiv*, VI, 81-93.

. . . is a law which God has placed in the nature of the thing. God is first of all and above all the one who works. With this we refute the sort of generation which the erring maintain and refer to as infection, in which the Arabs in their ignorance (before Islam) believed.

Ibn al-Khatīb, however, stepped forward as vigorous champion of the theory of infection. If he is asked, he says, how he can grant infection when the religious law denies it, he will answer, "The existence of infection stands firm through experience, research, mental perception, autopsy, and authentic knowledge of fact, and these are the materials of proof." He supports his stand by facts somewhat like those of Ibn Khātimah, though less striking and convincing, including various instances of immunity. One is of the pious Ibn Abū Madyan in the city Salé, who, believing in infection, had his establishment stocked with provisions and walled up with the numerous inhabitants inside; and while the city died off not one of those within was stricken. This is understandable in the light of present knowledge, as are the statements that isolated regions not cut through by roads and tent-dwelling nomadic Arabs of Africa remained untouched. But, though also possible, it is not so easy to accept the incident which he calls the most wonderful thing during this time, that several thousand Moslem prisoners of war in the dockyards of Seville were protected by God and kept free from the pest, while the city nearby was completely wiped out. He may feel this himself, for he begins his next sentence with "Certain is the news," and almost at once returns to the point about infection.

The sword of the pest, he says, raged among men, and God gave them into the power of a few lawyers, who fought against them with fetwas[26] as hostile heretics fought them with swords, and under the points of their reed pens were sacrificed the lives of so many persons that God alone knows their number; though the intention of the lawyers was pure, and they supported themselves by the letter of tradition.

But it belongs among evident principles that a demonstration evolved from tradition, if it is opposed to the perception of the mind and the evidence of the eyes, must necessarily be subjected to explanation or interpretation. And this, in the present instance, is exactly the idea of many who defend infection.

He gives one or two passages from the Holy Law which might admit a milder interpretation, but declares that it is not the task of medical science to discuss whether or not infection is in conformity with past interpretations of the Law. Everywhere the attitude of the people shows that they are beginning to take their stand against such argument as wickedness toward God and non-recognition of the value of the souls of Moslems.

Already, in fact, have pious people arisen in Africa, who retracted their earlier opinion, and formally proved by document that they withdrew from the earlier fetwa, since they consider their consciences burdened by the view that it is permitted them to surrender themselves to destruction.

So it is evident that in the western Moslem world the Black Death stirred up something of a revolt of medical science against theology.

Among the Christian writers, who are not forbidden to believe in infection, it is usually accepted almost without discussion. It is not mentioned by the University of Paris, Simon of Covino, Alfonso of Cordova, the Strasburg

[26] Decisions of the mufti, or lawyers.

Treasure of Wisdom, or in the tractate beginning *Is It from Divine Wrath*. The author of *The First Question Is, What Are the Diseases Now Current* gives the particular, near, extrinsic cause as the breathing in of the corrupt air which the suffering breathe out; the disciple of John Hake cautions that it will be well to refrain from associating with people, and that holding the breath is a good thing, while flight is to be recommended. The tractate *First about the Epidemic* says that the disease is contagious, and infects bystanders by breathing alone, or by long proximity, and in one copy of the treatise of John of Penna we find intercourse between people, in another the breath of those infected, given as among the means of entry of the disease into the body. Gentile of Foligno, who shows the influence of much study of Arabic works,[27] and is inclined to give more weight to corruption of the air than to infection, notes that the traces of the scourge remain for a long time and infect like a ferment. As a corollary he makes observations, strongly reminiscent of those of Galen, that it is dangerous to have dealings with those who suffer from the pestilence or with those who breathe the putrid matter, or to enter the house in which the fetor lies.

The practitioner of Montpellier has his own ideas on the subject of infection, and after all it is by the elimination of many theories that working hypotheses are established. His views are best expressed in his own words:

[27] Gentile carefully gives credit where it is due, quoting by name, and often by work and chapter, Haly Abbas, Avenzoar, Galen, Avicenna, the *Conciliator* of his admired master Peter of Abano, and others. It is interesting to note his use here of "seeds (*semina*) of disease," as a substitute for our germs. In doing so, he follows Haly Abbas, but I have seen the phrase in Galen also. His word for traces of the disease is *reliquiae*. See the beginning of the first chapter of his longer *consilium*.

Since therefore this epidemic, according to some, happens only by the air, only by breathing, only by conversation with the sick, more say that it kills because, by means of the air breathed in by the sick and then by the well standing near, the latter are stricken and killed, especially when the sick are in agony; and that not suddenly, but at intervals and gradually. But the greater strength of this epidemic and, as it were, instantaneous death is when the aërial spirit going out of the eyes of the sick strikes the eyes of the well person standing near and looking at the sick, especially when they are in agony; for then the poisonous nature of that member passes from one to the other, killing the other.[28] Whence whoever has seen the *Book on Mirrors* of Euclid about burning and concave and reflex mirrors will not wonder, but will grant that this epidemic can occur, and pass from sick to well, and the latter be killed naturally and in the nature of the case, and not miraculously; since a thing is miraculous when there is no reason or natural cause for its occurrence. But the aërial and subtle nature going forth and reflected from two mirrors, by means of the heat and brightness of the sun, immediately takes fire and, as it were, acts suddenly, contracting the diaphanous air by virtue of the brightness simply generated [?] from solar rays and mirrors; from which brightness buildings and houses and fortified places and trees, situated in that vicinity, are burned and destroyed; example of which may be had in the book of Euclid. Thus also by corruption of the air attack is made on human bodies, and more quickly on them than on any other anywhere because [*prope?*][29] of the first soft matter of which they are composed . . .

A little later on, at the conclusion of his description of the course of the disease in the body, he elaborates somewhat on this means of infection.

Then the sick die soon afterward; and sometimes the brain expels this windy and poisonous material through the concave

[28] As sound a fourteenth century medical authority as Guy of Chauliac thought the plague could be contracted visually. See above, p. 3.

[29] The Latin is poor.

optic nerves to the eyes, and then the sick person is in agony, holding his eyes as if they could not be moved from place to place, and there the first ventosity receives a marvelous property, in that, thus standing and permanent, that toxic spirit is continually being made, and seeks a dwelling-place in some nature into which it can enter and lie quiet. And if any well person looks upon that visible spirit, he receives the attack of the pestilential disease, and the person is poisoned more quickly than by inhaling the air of the sick man, because that diaphanous poison penetrates more quickly than the heavy air.

Another instance of this is the basilisk, from whose eyes passes a visible, aërial, and venomous spirit to the eye of the person looking upon it, killing him. So that the way to take a basilisk is to cover the eyes and hold a light over the head; the creature gazes at the light, and may be killed. Mention of the basilisk occurs in two other of our tractates, but the allusions are of a skeptical nature. Ibn Khātimah observes that when a well has been shut up so long that the air becomes sufficiently corrupt to put out a light lowered into it, the ignorant say it was extinguished by a basilisk lurking there. And *Is It from Divine Wrath* goes further, declaring that it has been proven by a very great number of experiments upon wells shut up for a long time that the air within becomes poisonous to human nature: for when they are opened, sometimes the first who enters is suffocated; and the vulgar and ignorant think it is the basilisk lurking there.

In Ibn al-Khatīb's treatise we find recognition of the fact that the pneumonic type of plague is the most infectious, and a statement of the development of immunity against pestilence on the part of persons who have lived in pestilential localities until they have apparently assimilated themselves to it. Dionysius Secundus Colle notes that almost all those who do service like that of orderlies in hospitals and other public places seem immune from pestilence, for poison fights against poison and drives it out.

Besides the astronomical causes of the epidemic, the corruption of the air and infection, individual factors enter into the contraction of the disease, especially a person's susceptibility to it and the preparation of the body to withstand or receive the infection.[30]

Those persons are particularly susceptible, according to the University of Paris, whose bodies are replete with humors. Ibn Khātimah extends this somewhat to persons of hot, moist temperament, if young and corpulent, especially young women of strong sensuality and passions; persons so disposed should be very careful about food, drink, and sleep. Ibn al-Khatīb explains that women and children are attacked in a more horrible form because of the humors which are preponderant and are connected with warmth. He goes at length into the question of predisposition: it is the preparedness of one thing for the reception of another by reason of the analogy or similarity which exists between them. If the individual temperament is like that of the poisonous power hurling itself upon it, and is inclined to it without resistance, as quicksilver throws itself upon gold and amalgamates itself with it, then this power presses into that of the individual temperament and unites with it, spreading into the juices and humors with the spread of the vital principle, corrupting this principle as poisons do. But the individual temperament, if unlike the temperament of the poison, withstands it; and sometimes such an antithesis, though lacking by nature, may be cre-

[30] Galen stresses the condition of the body as the chief factor in contracting the disease or throwing off infection. "Subjiciatur ergo pro exemplo in ambiente aere invehi quaedam pestilentiae semina . . . ea vel nullo modo, vel exigua labe infici, ut facile ad habitum naturae recurrant." VII, 291–292.

ated by the physician. An illustration of these facts may be found in a wick brought near the flame of a lamp: the wick which has recently been lighted takes fire at once, but the dry wick takes longer, and the damp, only after long waiting and effort, if at all; these are the pictures of persons completely, slightly, and not at all predisposed. John of Penna says that in the pestilence the more robust perish, the greater disturbance having been occasioned in their strong natures; and of the rest, more women and boys—not that such matter is more abundant or more acute in them, but that it is more contrary to their nature. When he also includes among the especially susceptible the decrepit and the old, and, in the Wiesbaden copy of his work, girls reaching maturity, he has embraced the whole human race.

Simon of Covino limits more narrowly, and more in accord with what might be expected, those whom the plague takes by preference: "But the fragile nature succumbs, nor can stand for long, however the hand of physicians may help. He who has been poorly nourished, with slight strength of foods, falls stricken by the least breath of the scourge. And next Saturn's throng, the multitude of paupers, give way to grateful death, since living is death for such. After them the lunar[31] perish, and the mercurial. And thus the weaker yield in the first rank; afterwards others finally become victims of the same pestilence. But cruel fate spared the prince, and the noble, generous soldiers, or judges. Rarely do such fall, since to them is given a delightful life, which idle glory praises." Which seems still to leave open the question of the effect of the Black Death on men of learning.

Consideration of the preparedness of the body to resist

[31] Those of moderate condition.

or succumb to the disease properly belongs in what forms a substantial portion of every treatise, the section on protection or preservation against the plague. Preventive steps must be taken against each set of causes, or, as the tractate *The First Question Is, What Are the Diseases Now Current* puts it, precautions are to be taken against the pestilence by resisting those things upon which it depends. People may be puzzled as to how to resist the heavenly cause, but the way it is to be done is by wearing a little image of a lion carved on gold alone,[32] in the very lucky sign of the lion through the aspect of the benevolent planets. None of our other authors gives so concrete a remedy against the heavenly cause: the usual recommendation is prayer, and Simon of Covino urges that to this divine hope be added the aid of nature, obtained by following her advice. This turns out, however, to be flight from the danger as quickly as possible.

Flight is stressed by all the writers as the first and best way of avoiding both corruption of the air and infection. If it is impossible for people to leave the place where they live, they should at least, the University of Paris advises, choose a place of residence low, so as to be sheltered from corrupt vapors, far from marshy and fetid places and stagnant waters, with windows open to the north.[33] Those

[32] The Latin is "in auro solo," which Sudhoff suggests may originally have been "in auro solis." The whole passage is: "Qualiter igitur caelesti causae obviabitur? qualiter aliter quam secum gestando sigillum ymaginis leonis sculptae in auro solo in leone bene fortunato per aspectum benivolorum planetarum, sicut docet Thebit et plures ex summis astrologorum." Much like a favorite image of Peter of Abano. See Thorndike, *History of Magic,* II, 899.

[33] Avicenna's directions do not seem to be inspired by practical intelligence: pestilential air should be fought by making the house fresh with cool, moist things, or by heating it and fighting it with its opposites, good odors, and by using fans to drive out impure

facing south should be kept closed and fastened, at least till after the sun has risen or a fire has been made. The windows should be glazed or covered with waxed cloth to partly keep out the air, except when the north wind is blowing pure and clear about midday. Dwellings in woods are undesirable, though those are better which are clear of the woods on the north. Ibn Khātimah considers cities on the coast, especially when the sea is to the south, more susceptible than others, because the rays of the sun and other stars are reflected from the water mirror to the cities, making them warmer and damper. Likewise cities facing the south pole are in more danger than those facing the north, particularly when unprotected on the south side; while those facing east or west occupy a middle position, west being worse because damper; and those facing obliquely should be judged accordingly. Cities on plains are especially susceptible, and even more so if in a depression or lowlands near valleys, standing waters, or fields, being without power of resistance because of the reflections of the sun's rays, the dampness, and the immobility of the air. The opposite is true of mountain cities because of their coldness and dryness. Houses, too, should face north, and people should avoid the sun, warm winds, stoves, and

---

air. Its corruption often comes from the earth, in which case people should stay in their rooms and seek high shelters shielded from the wind; or it may begin in the air itself, because of unknown celestial influences, whereupon refuge should be sought in subterranean dwellings, or in those surrounded on all sides by walls and caverns. There should be fumigations, too, as of cypress, frankincense, myrtle, rose, and sandal. *Canon,* p. 129; Lib. I, Fen. III, Doct. V, Cap. 1. Earlier he has paid a glowing tribute to the north wind, which "strengthens and hardens, checks those things which are manifestly fluid, closes pores, strengthens the digestion, constricts the stomach, provokes the flow of urine, and renders healthful, putrid, pestilential air. *Canon,* p. 64; Lib. I, Fen. II, Doct. II, Cap 10.

everything which generates heat; the reason is clear to members of the profession. John of Penna in one *consilium* advises living in mountain and purer air, and in solitary places on account of the danger of inhaling others' breaths and of going into crowds; in the other he makes flight from the smitten the first rule, and as to the direction of the wind, counsels that it should come from the opposite direction from the pestilential region.

Whether or not a choice can be made of place of habitation, the air should be rectified by burning in the houses dry and odoriferous woods, such as juniper and ash, vine and rosemary, young oak and pine. In winter aromatic things ought to be burned, wood of aloes, amber, and musk for the rich, and for the poor, costus, cypress, laurel, mastic, and the like; and on the fire are to be thrown troches for which various recipes are given. A comparatively simple one from the tractate *The First Question Is, What Are the Diseases Now Current* calls for one ounce each of choice storax, calamite, and wood of aloes mixed in a mortar with a pestle, rose water of Damascus to be added, and from this mass small oblong troches made and dried in the shade; by throwing one of them on the coals a whole room will be fumigated. In summer the house is to be filled with cool, pleasant-smelling plants and flowers, and sprinkled with vinegar and rose water. Ibn Khātimah advises sprinkling oneself with them, rubbing them on the face and hands, and smelling often of them; he includes such things as citron and lemon, roses and violets. Most of the tractates prescribe carrying in the hand and frequently smelling an amber or smelling apple, the recipes varying in the number and costliness of ingredients. The University of Paris gives four: the first, for the king and queen, is an apple of pure and finest amber, for amber, largely on account of its aroma, greatly cheers and invigorates, strengthening the principal organs and increasing the vital principle. But since amber is dear, the other three recipes are given; the simplest, that of the ninth-century Arab physician, John Mesuë, prescribing equal parts of black pepper, and red and white sandal, two parts of roses, half a part of camphor, and four parts of bol armeniac. All but the camphor are to be ground very fine, sifted and shaken, pounded during a week with rose water, then the camphor mixed with them, and the apples made with a paste of gum arabic and rose water. Gentile of Foligno suggests for the poor the smelling of any sort of odoriferous herbs; and in his usual practical vein, while speaking of fires, he says that something which is useful for everyone is building a fire in the streets with material gotten by cleaning dirty houses and cities.[34]

Personal prophylaxis[35] is somewhat hampered by the

[34] Galen, XIV, 281, attributes to Hippocrates, at the time of the great Athenian pestilence, the instructions to build fires throughout the city, cast on them flowers and wreaths, and to purify the air with pungent unguents. Avicenna does not mention fires, but does urge suffumigation; he stresses chiefly rectification of the air by sprinkling with cooling and odoriferous substances. See above, note 33, and also *Canon,* p. 807; Lib. IV, Fen. I, Doct. IV., Cap. 4.

[35] Of our two earlier authorities, both express themselves on the subject of bathing and exercise in time of pestilence. Galen, VII, 286, gives warning that those who are bathed rashly are prompt to take fever; and they who have gone down into a warm bath of sweet waters cause the corrupt foods to be carried through the whole body. The latter result also follows too violent motions, or standing in the summer sun. But, though he may not like exercise in too violent a form, he favors it, saying a little earlier (VII, 279) that it has been found by long experience that those who have left off the habit of exercise are carried off by many diseases, and especially fevers.

Avicenna cautions against bathing in cases of fever or imposthumes because if prolonged it weakens the heart, causes faintness and nausea, makes quiet humors move, prepares them for putrefaction, starts them toward the receptacles of weak members; therefore

general idea that a hot bath, or, to some extent, any sort of bath, by opening the pores admits to the body the corrupt air, so that it is better to bathe less often than usual, if at all, only in tepid water, and not to stay too long in the bath. But the face and hands should be washed frequently in vinegar and water, rose water, a decoction of rosemary, or warm wine. In one manuscript of the tractate *First about the Epidemic* washing the mouth and nostrils with vinegar is recommended, in another, though this is rather to aid digestion, gargling for a long time with a preparation of softened sage, followed by slow and prolonged chewing of certain things, and swallowing them.

Exercise is frowned upon as introducing into the body a greater amount of air, and consequently of the poison with which it is filled. John of Penna discountenances bodily or fatiguing labor, and Ibn Khātimah gives what must have been a popular prescription when he advises resting as much as possible. Moderate exercise in good weather may be indulged in if one is accustomed to it, and both it and the bath are permitted if they aid in preventing constipation.

Active steps against the poison should be taken, and a dozen or more antidotes are given. One which is almost universal consists of a fig or two with several filberts and some rue before breakfast on a fasting stomach; this is the first of the set of directions received from John Hake, bishop of Freising, followed by the rash promise that he who takes it will not contract the plague. The same claim is quoted from the ancients by other tracts in connection

from them occur imposthumes in hidden parts of the members and on their surfaces. He gives directions for kinds and times of exercise, and for massage. *Canon,* pp. 112–115; Lib. I, Fen. III, Doct. II., Caps. 1–6.

with the often prescribed pills of aloes, myrrh, and saffron; and we find several combinations which seem to be more original, such as a powder of black pepper and cummin which the practitioner of Montpellier suggests as an ingredient for all foods, condiments, and sauces. But the most respected of all antidotes are the four time-hallowed remedies against poison; theriac, mithridate, bol Armeniac, and terra sigillata.[36] Though bol Armeniac seems to be preferred by the University of Paris, theriac is usually the most highly favored of the four, if it be properly compounded and aged. According to the medical faculty of Paris, it should be ten years old, be taken once or twice a month in the quantity of about a dram mixed with two ounces of good aromatic wine, after the system has been thoroughly cleansed, and for about nine hours, or until it is entirely digested, no food should be taken. The esteem in which it was held was so great that the word was a synonym for antidote: onions, for instance, are "theriac against poison of every sort." Onions should, however, be eaten only by strong men, accustomed to coarse food, drinking little or no wine, and living near bad waters, and then

[36] Theriac and mithridate were both electuaries, supposed to fight against poison by the immunity produced by gradually introducing into the system small quantities of poison of the same sort. Theriac, from which comes our word treacle, means literally a remedy for snakebite, and was compounded of more than sixty different ingredients, the main one of which was often chopped-up bits of snakes. Mithridate was named from Mithridates VI (132–63 B.C.), king of Pontos, its supposed inventor. Bol Armeniac (Armenian bole) and terra sigillata were preparations of clay highly esteemed for their medicinal value. For further information about them, see Lynn Thorndike, *History of Magic and Experimental Science*, 2 Vol., New York, 1923; and George Sarton, *Introduction to the History of Science*, Vol. I, Baltimore, 1927. Both Galen and Avicenna advise their use against pestilence.

especially in winter, since, through warming the cold humors of the body and removing the gross ones, they develop diseases to which the body has a tendency. The Bourges manuscript of *First about the Epidemic* gives as a substitute, if approved theriac cannot be obtained, a mixture of rhubarb and spikenard to be taken in water of endive and white wine. Gentile of Foligno says the theriac should be a year old, and since the poor cannot obtain it they should use instead white horehound or hyssop and scabious or ortolan; for children under thirteen, who are too young to take theriac internally, it should be used to anoint the heart, stomach, and nostrils. Further virtues of the four classic remedies against poison are that they strengthen the heart and cheer the spirits. Other antidotes are given, both herbal prescriptions and certain uses of precious stones. Powdered emerald is excellent if taken internally, for, Gentile of Foligno tells us, its efficacy against poison is so great that it has been found that if a toad looks at it its eyes crack; if it be put on the table it weakens a poison; and if in the mouth it lessens the power of infection. This information Gentile obtained from his admired master, Peter of Abano,[37] to whom he is also indebted for the ninth prescription against the poison which he has already given: it is found, he says, in the book of the Persian kings, and strengthened by the authority of the *Conciliator*, that if there be inscribed on an amethyst a man bowing, girded with a serpent whose head he holds in his right hand and tail in his left, and this stone be put in a gold ring having under the stone the

[37] For a discussion of the *De venenis* of Peter of Abano, see Thorndike, *History of Magic*, Vol. II, ch. 70. See also tale of toad and emerald in Albertus Magnus, in the same volume, pp. 546–547.

root of the serpent, and the ring be worn, it will preserve from all poison.[38]

After taking measures against heavenly causes, corruption of the air, and infection, much attention should be given to the preparation of the body to withstand disease. Then, as today, purgation was the essential thing, and for this purpose first natural means, food, exercise, and massage, were to be employed; then, if they were not efficacious, laxative medicines such as a decoction of prunes or cassia; or suppositories or the clyster. Diuretics were to be used for the urinal passages. Very important for cleansing the body was the drawing off of impure or superfluous blood; against the necessity of bleeding no voice was raised, but as to time, place, and manner there was diversity of opinion.[39] The references to bleeding for prevention are rather cursory in most of the tractates, but Ibn Khātimah speaks with the ring of conviction based on theory and practice: the superabundance of blood which makes bleeding extraordinarily necessary during the pestilence is due to the heat and moisture of the air resulting from its change in constitution. This heat greatly increases the natural heat of the body, thereby transforming the chymus into blood, which in turn becomes overheated and corrupts

[38] Long *consilium.* The portions cited may be found in the portion of the *consilium* published by A. Philippe, *La Peste Noire,* Paris, 1853.

[39] Galen, XIX, 524, says that at a time of serious pestilence in Asia he escaped by having two pounds of blood drawn from his leg. Others who had moderate headaches and sore throats caused these symptoms to disappear by bleeding. Avicenna, *Canon,* pp. 145–150, Lib. I, Fen. IV, Cap. 20, says that bleeding should be practised only when one has contracted or is contracting a disease: "Phlebotomia est evacuatio universalis, quae multitudinem evacuat. Multitudo autem est augmentum humorum, super aequalitatem ipsorum in venis."

the natural temperament of the heart: it can happen that the blood, through its abundance and moisture, suddenly extinguishes the natural heat and it goes out as does a lamp when its wick is submerged in oil. Life can maintain itself until the excess of blood overcomes it. Consequently, though ordinarily the time of the year is to be considered in bleeding, such is not now the case, for "this disease changes all times of the year into one time, all diseases into one disease." People who have never been bled before, and even children, now have it done repeatedly, and experience no ill effects, but only beneficial ones. Ibn Khātimah has seen people during the pestilence have as much as eight pounds of blood drawn, though the usual amount is about five pounds. He had himself at first hesitated to prescribe it on account of the time of the year, but he did try it, and the results were such that he had continued it ever since. "After people learned this and saw its effects, they began to have bleeding done for themselves, without medical prescription, several times a month, without consideration or fear, without feeling harm or weakness, without contracting sickness in consequence. God has as a result of this enlightenment caused great effects to follow in those whose sparing was ordained."

Diet is stressed in the treatises as one of the most important factors in keeping the body in good condition.[40]

[40] Galen, VII, 284–287, divides into two classes foods that are bad: 1. that bad by nature, as garlic, onions, cress, leeks, cabbage, ocimum [a sort of clover], urtica [a nettle], and others which are called *agrestia*, like charlock; 2. that good by nature, such as barley and wheat and other grains, but spoiled by age, bad storage, or rust; also bad are glutinous and fat foods, which obstruct the channels of the body.

Avicenna, *Canon*, pp. 115–121, Lib. I, Fen. III, Doct. II, Caps. 7–8, advises fruits such as figs, sweet ripe grapes, and dates in the regions where they are grown; meats such as kids, suckling calves, and

Their readers are told, as most of us also have been at various times, to chew slowly and rise from the table hungry; fasting is to be avoided as well as eating to excess; and the practitioner of Montpellier warns that repletion of drink is worse than repletion of food. Meats as a general thing are inadvisable, as are foods which spoil easily, such as milk and fish; if meat is eaten it should be tender and digestible, such as fowl, veal, lamb, and kid, and, the University of Paris says, roasted rather than boiled; if fish, it should be from fresh waters and recently caught. A little cheese may be taken after meals to aid digestion, but not otherwise; mushrooms are forbidden by Ibn Khātimah, and pastries and fried things by several authors. Eggs may remain part of the diet, though with restrictions: Ibn Khātimah says he has nothing against hen's eggs dressed with vinegar without garlic, while *First about the Epidemic* discriminates against hard-boiled eggs.

As regards bread, it should be made of good flour and carefully baked: the sort to which one has been accustomed from childhood is to be preferred. Ibn Khātimah is in accord with modern dietetics in saying that coarse bread is better than fine, and black better than coarse, though the reason he gives is that it clears the alimentary canal of chyme. He also recommends eating vegetables and fruits for their laxative properties, but most of our writers are inclined to frown upon them as being cool and moist, and not so easily digested. Decided differences of opinion appear, the usual advice being to avoid all fruits except certain named ones, the exceptions varying with each author, and both leguminous vegetables and vegetables of

---

yearling lambs; and wheat of good harvest and well stored. Watery foods corrupt the blood. Choleric or bitter vegetables are sometimes healthful in winter.

other sorts, each with named exceptions which do not tally in the different tractates. For instance, Paris cautions against eating lettuce, which is recommended by Ibn Khātimah and Gentile of Foligno; and cabbage in winter is one of the few vegetables permitted by *First about the Epidemic,* but is, with eggplant and garlic, expressly forbidden by Ibn Khātimah. The latter permits other vegetables, thereby showing more latitude than most of his contemporaries. The favored fruits are figs, dates, raisins, and pomegranates; nuts are included with fruits, the one universally recommended being the filbert. Sauces are declared to be necessary, in winter to be seasoned with pepper and other spices, in summer with vinegar, verjuice, or sour wine.

The best drink, according to the Christian authorities, is old, light, aromatic, white wine mixed with water—boiled water, Gentile of Foligno directs. For those who are not accustomed to wine Paris recommends mixing vinegar with water. Water for drinking should be pure, clear spring water, or, if from a river, should be taken where there is a strong current flowing over a rocky, not an earthy, bottom. If the drinking water is not above suspicion it should be purified by boiling or by distillation in an alembic. Ibn al-Khatīb does not concern himself with food and drink, but the orthodox Ibn Khātimah forbids even the permitted wine, and of course the forbidden, because it heats and increases the blood. The University of Paris has already indulged in a brief argument on that score, with the observation that if some of the ancients forbade wine it was probably because its use was less widespread among them "than it is today with us," or because their wines were stronger, or because in treating a pestilential disease they considered it as arising from heat rather than corruption.[41] Another difference as regards personal habits, showing the difference in point of view between Christian and Mohammedan, is seen in the requirement by the former of absolute chastity in time of pestilence, by the latter of chastity to a reasonable extent.

Sleep immediately after eating is bad, as in fact is any sleep during the day; the normal night sleep is best, without prolonging it, for that corrupts the mixtures, or shortening it, for that burns them out and destroys the vital principle. Gentile of Foligno gives directions regarding position for sleep which strikingly recall those of present-day digestive faddists: [42] sleep should be begun when lying not on the back but on the right side, and later the position should be shifted to the left side; it is a good thing to lie on the stomach first, since digestion is thereby aided by the retention, compression, and consequent increase of the natural heat. His reasons for the change from side to side have to do with the heat of the liver, and the objection to lying on the back is that it causes the flow of the superfluities to the palate and nostrils, then back to the brain, submerging the memory; from this are frequently produced nightmare, imposthumes, frenzy.

The last injunctions touching the condition of the body consequent upon manner of life deal with what are usually termed accidents of the soul. It must be noted, the University of Paris observes, that diseases of the body

[41] Avicenna, *Canon,* p. 115; Lib. I, Fen. III, Doct. II, Cap. 7, advises wine of good savor, but in the next chapter discourages wine-drinking as being harmful to liver, brain, and nerves, and leading to nervous troubles, madness, and sudden death.

[42] He gives as his authority Haly Abbas; directions of a similar sort may be found in Avicenna, *Canon,* p. 121; Lib. I, Fen. III, Doct. II, Cap. 9.

often arise from accidents of the soul; or, as Gentile of Foligno puts it, the body is often moved in various ways by imagination alone, for imagination works havoc. People should guard against fear, worry, weeping, speaking ill of others, cogitation, and wakefulness; and especially to be avoided are wrath (which, in the opinion of Gentile, overheats the members), and sadness, which cools the body, dulls the intelligence, and deadens the spirit. Ibn Khātimah goes even farther, considering the latter one of the main causes of pestilence. Worst at escaping it, in his experience, are the intellectually gifted; best at doing so are idiots and the dull. Which may cast a further helpful ray of light on the effect of the Black Death upon men of learning.

In treating the cure of the disease the various tractates give discussions of its nature, clear statements of its symptoms, and graphic descriptions of its course in the body,[43]

[43] Galen and Avicenna do not appear to have been very helpful to fourteenth-century physicians in specific details of symptoms and cure of plague; this may be explained by the fact that Galen left Rome just before the great pestilence of the second century A.D. reached it, and that there was no notable outbreak of pestilence during Avicenna's lifetime. Galen's definition of pestilence is "a disease arising from corruption of the air, and attacking all or a great many of the people of a country, resulting in the death of many." (Vol. 19, p. 391.) The symptoms recounted by him and Avicenna are those of fever, except for the inclusion of imposthumes and ulcers; and the cure the usual prescriptions for cleansing the body by bleeding and purgation, care about diet, and, besides, directions for rectifying the air. Galen makes the rather remarkable statement that it seemed to him easy to cure those who were suffering from the pest by drying out and purging their bodies. Then an eruption appeared over the body, sometimes ulcerous but always dry, which was nothing but what was left of the blood corrupted by the fever: ashes, as it were. These required no treatment, but healed of themselves. Sometimes these occurred on the lungs, but were healed by drying them up; anyone who should follow this method would be able to save many who were spitting

the last rivaling in dramatic interest present-day accounts of the struggles taking place within us between white corpuscles of the blood and invading bacteria.

The nature of the pestilence is considered by the Spanish-Arabic writers, Ibn Khātimah and Ibn al-Khatīb, and by the Italian Christians, Gentile of Foligno and John of Penna. After definitions of pestilence and epidemic in general, Ibn Khātimah characterizes this special malady then current in Almeria as "a malignant fever which gets control in consequence of the corruption of the temper of the heart, a corruption arising from the change of the air from its natural condition to heat and dampness." Later he points out that this pest-fever is different from all other fevers, and, as it were, exactly opposite to them, as in the others the fever goes from the diseased part to the heart, here from the heart to the body.[44] Ibn al-Khatīb's definition is somewhat fuller: "an acute disease, accompanied by fever in its origin, poisonous in its material, which primarily reaches the vital principle by means of the air, spreads in the veins and corrupts the blood, and changes certain humors into a poisonous character, whence follow fever and bloodspitting and breaking out in exantheme of a pestilential sort." Gentile of Foligno has, "after much careful work and dif-

---

blood from their lung, as Galen had healed them. X, 367–368. Some authorities think that the pestilence of Galen's time was smallpox rather than bubonic plague, but the bloodspitting seems to point to the latter.

[44] "The pestilence is a general disease which befalls mankind, usually running a fatal course, and arising out of a universal cause." Epidemics come through the air, not through food and drink; do not afflict a man in consequence of his own deeds, but from without; and befall great masses of men. Medicine is defined as "an art which through research and experiment has arisen with the object of maintaining the natural temperament, and of restoring it to him who has lost it." *Archiv,* XIX, 31–32.

ficult experiments, found the immediate particular cause to be poisonous materials which burden the heart and lung, the attack of which is not from excess of degrees of primary qualities, but through the peculiar character of the poison. . . . There seems moreover to be connected with this principal disease a certain generation of worms."[45] Gentile's theory is attacked by John of Penna, who declares that it is impossible, while life still remains in the body, for such material to be generated of poisonous property as Gentile claims. His own contention is that the disease is caused by coleric matter located in the channels of combustion, like the matter of anthrax, mixed with blood in the veins. Closely connected with this coleric matter are a peculiar heat and corruption, and their point of attack is particularly the chest and heart.[46] This is the version of John of Penna's views given by the writer of the Bourges manuscript; in the Wiesbaden codex they appear in slightly different form. There the pestilence is stated to be, not a pestilential fever like those described by medical writers, but pestilence with fever, and to kill, not by excess or crudity of febrile matter, since it kills most on the first, second, or third day, but by its insidious poisonousness: for it is anthracal pest, or pestiferous anthrax.[47] In this treatise he has already defined it as glandular pestilence.

The symptoms[48] are fever, pain in the side or chest, coughing, difficulty in breathing, with the breath short and

[45] *Archiv,* V, 84.
[46] *Archiv,* V, 341–342.
[47] *Archiv,* XVI, 164–165.
[48] Full descriptions of the symptoms are given by Ibn Khātimah, Gentile of Foligno, and Dionysius Colle; briefer ones by Ibn al-Khatīb, John of Penna, Alfonso of Cordova, and the authors of the tractates *First about the Epidemic* and *The First Question Is, What Are the Diseases Now Current.*

close and a pulse of the same kind; and vomiting of blood, or appearance of buboes behind the ears, in the armpits, in the inguinary region, or, also, on other members. A distinction is made between the pneumonic and the bubonic types, and it is observed that the former is the more infectious and the more speedily fatal: Ibn Khātimah says that he has seen only one man recover from it, and Ibn al-Khatīb states that it cannot be cured, and that it is more infectious than the other. Ibn Khātimah distinguishes a third variety, that accompanied by abscesses, while Dionysius Colle mentions gangrene of the feet as one of the ways in which people are affected. The indications of the onset of the disease noted by Simon of Covino are pallor, a bitter taste in the mouth, and the darkening of ruddy complexions; he also comments upon what physicians of today describe as an expression of intense anxiety on the face of a patient in the early stages of bubonic plague: "Hardly is there one whose face does not pale; merely in the expression of women and men can the dire writing be read, and the imminent calamity of death on the way is betokened by pallid signs; before the day death is seen seated in the face." Alfonso of Cordova warns that an incipient imposthume makes itself known by prickling of the skin; if he who feels that prickling wishes to know whether it is a true imposthume, let him cough; if thereupon sharp pains occur in the same place, it is a true imposthume. The author of *First about the Epidemic* describes the appearance of the bubo thus: "This is the sign of this disease. For in the inguin or the side, under the arms or on the neck, or in some other place a little something begins to swell and then grows."

The course of the disease in the body is traced by Ibn Khātimah, whose point of departure is the entry of the corrupt air into the system; by Ibn al-Khatīb, on the basis of the theory of infection; and by the practitioner of Montpellier, from an astrological starting-point to a conclusion of ocular infection. Several other writers, as John of Penna and the author of *Is It from Divine Vengeance,* devote a few lines to the subject.

The two Arabian physicians first trace the progress of the malady before it takes root in the system: Ibn Khātimah explains that man's life depends upon the natural heat of the body, the source of physical and spiritual strength, which comes from the heart. Thence the heat is driven through the arteries which arise in its left cavity into all the arteries. This heat cannot fulfil its task unless it remains natural and uniform, and so God has given man breath, with larynx and lung as organs of breathing. They are fastened to the chest in order to expand with it, and when a man inhales the lung expands and takes in the air, which is driven into its tubes and cavities, to the left cavity of the heart, and from there to the lung. So far the air has been purified, the heat of the heart moderated, and the boiling stopped; and in exhaling the lung presses out the air laden with accumulated vapors. Were the heat not moderated by the air the juices of the body would be burned away and life destroyed, while should the damp humors of the heart remain they would extinguish its heat and destroy the vital principle, as too much smoke extinguishes fire. But the breathing apparatus requires undefiled air; when it is changed in its entirety "it makes the vital principle corrupt with its own corruption," replaces the natural heat of the body with a foreign heat, and allows the vapors to overpower the vital principle. The fever accompanying the pestilence is the opposite of other fevers in that it attacks the heart first, then the other organs;

whereupon ensue noteworthy symptoms, as when disorder breaks out in a city in consequence of the failure of the commander. For the heart cares for order in the body as the king in his city, the master in the house; and when disorder occurs in other organs the heart, even though it may itself be incidentally affected, takes control and drives out the disease, to the extent of controlling medical means used. But when the disorder occurs in the heart itself it cannot take care of itself or the other members, and it is very doubtful what may be the results of medical treatment. The best thing is to try to strengthen the heart, and the most efficacious means of doing so is by venesection, which removes the excess of corrupt blood, and also, by emptying the veins, clears the way for the vital principle.

Ibn al-Khatīb makes the disease appear in the body not from corruption of the air but through predisposition, which is rare, or through transmission and infection, which is the usual way. Thus the vital principle becomes involved after more or less resistance, grows heated, fever sets in, and the foreign heat presses into the arteries and becomes general. The humors in the veins are corrupted, in the blood sets in an agitation to drive out the destructive humors, and nature hastens to thrust them out according to the strength imparted to her by God. If she is strong, and is assisted by certain relations of earth and heaven, as is taught by masters of the art, she succeeds by means of crises, secretions out of the canals of urine, excrement, perspiration, nose-bleed and other hemorrhages, and recovery begins.

If the disease takes root the heart, according to Ibn Khātimah, or nature, according to Ibn al-Khatīb, next attempts to get rid of the poison by sending it to some outlet, where effort is made to localize it. The outlets are the places meant by nature to receive the superfluous matter from the noble organs, corresponding to ashpits in the house: the cavities behind the ears for the brain, the armpits for the heart, and the groin for the liver. Ibn Khātimah reveals further that if the harmful mass is thin it mounts to the head and is driven to the outlet of the brain; if it is thick and heavy it falls to the lower parts, and the liver drives it to its outlet; if medium, it goes to the heart's places of elimination. Then buboes are formed, and in case that all goes well it may happen, Ibn al-Khatīb says, that the contagion may be limited to these places, the buboes will ripen, discharge the poisonous matter, and a cure be effected.

But if the body is not strong and its humors are not sound, the harmful mass, instead of reaching an outlet, may attack some weak organ which lies in the way: this is most likely to be the lung, "a tender organ of thin tissue," both on account of its nearness to the heart which, like one who is heavy-laden, seeks to free itself of its burden wherever it can, and because of the lung's susceptibility, passivity, thin structure, thin blood, and uninterrupted movement. For the last three reasons, which are not conducive to cicatrization, the lung cannot localize the blood and is usually torn to pieces by it, if we believe Ibn Khātimah; Ibn al-Khatīb's view is that the lung becomes ulcered, symptoms of phthisis[49] appear in it, the neighboring parts of the breast are sympathetically involved in the corruption, and vomiting of blood ensues. Sometimes, Ibn al-Khatīb continues, there still exists in nature a remnant of power of resistance, and the poisons are cast to heart, liver,

[49] This phthisis, Ibn Khātimah says, is a sign that the lung is localizing the mixture, and rarely occurs.

brain, or other parts, after they had almost fastened themselves upon the lung and had already evinced a few of the above symptoms; but then there occurs a relapse, the matters turn—after the lung has exhausted its strength—back to their center, the power of the poison becomes sharpened, it wins the mastery, overpowers the vital principle, and quenches it.

Ibn Khātimah mentions other courses which the malady may take besides forming buboes or attacking the lung. If it attacks the liver, a hot swelling arises; if the diaphragm, pain in the body; if the intercostal muscles, stabbing pains in the side; if the palate, sore throat; if the brain, acute ulcer and headache. The mixture can be driven to two or more organs according to its quantity and their receptivity. If none of the inner organs is weak, and the mixture is especially strong, then it is driven to the surface of the body and forms boils, usually on the back and neck, rarely on the extremities. These are the sorts of the disease; those of most frequent occurrence are the two first, buboes and bloodspitting, and the third, abscesses.

The practitioner of Montpellier starts his exposition of the progress of the disease in the body with the dominance of Saturn, since because of that planet's frigidity growing things are damp and viscous, and, being eaten so, are badly digested in the stomach, and in the liver the blood is consequently diluted, weakened, and poisoned. From this is ordinarily generated the windy imposthume, and that is why most of these imposthumes are developed in the right side rather than the left. If the imposthume develops next to the liver, the pulse is frequently weak and slow. The humidity and poisonousness may be located elsewhere, or ascend thence through the organic vein to the brain, sometimes going to the lung and lingering there in the canals of the lung: the movement of the latter ceases and it cannot draw air in and out above the heart to cool it. Then the heart is heated and pestilential fever sets in, from which are set in motion humors ending in imposthume, which may be external or internal, filled with humors or with air; next the brain, in sympathy with the heart on account of its movement and spongy texture, draws its windy and poisonous humidity from the lung, sending it sometimes through the ears. Then is felt a great roaring as if a door is being broken; it is the first accumulation of air which cannot escape. Or the brain may send the material to the eyes, which it is likely to do more quickly on account of the diaphanous nature which it seeks, since the air is diaphanous. That is when the poisonous material is likely to escape into the eye of the beholder and infect him with the disease.

With a brevity which is welcome after the practitioner's diffuseness, John of Penna merely observes that the corrupt material is driven to the glands which are the proper emunctories, causing fever or intrinsic imposthumes, most of which are in the breast, or extrinsic ones wherever they may occur. *Is It from Divine Vengeance* confines itself to the cryptic statement that the poisonous matter, which may be taken in at any time of the year, may lie hid in bodies till it penetrates and finally perforates the heart's little dwelling, and thus, before men feel the attack, they become filled with the deadly stuff, till it penetrates that treasure house of life, namely the heart. And when this happens the infected die at once.

In most of the tractates the cure of the pestilence does not play a large part, since their authors are in full agreement with Galen that the crisis of a pestilential disease tends more to death than to recovery. This axiom is quoted by Master Albert, who, nevertheless, in view of the fact that the present malady is not only pestilential but very contagious, devotes the whole of his brief treatise to prescriptions for cure of the smitten. Only three of the other physicians differentiate appreciably between prevention and cure: they are Gentile of Foligno, John of Penna, and Ibn Khātimah, the last of whom is far superior in this field to his contemporaries. The University of Paris, whose compendium shows little first-hand acquaintance with the epidemic, says nothing of its symptoms or its course in the body, and combines prescriptions for curative remedies with those for preventive. Ibn al-Khatīb gives practically nothing about the cure, Alfonso of Cordova limits his advice along that line to instructions for blood-letting, and Dionysius Secundus Colle devotes himself almost entirely to an account of what people did to avoid the plague. Simon of Covino describes in a fashion more poetic than useful measures advocated by physicians, who, he remarks, almost all died; and the practitioner of Montpellier discusses only the causes of the pestilence, its course in the body, and means of prevention. Causes, course, and prevention also form the subject matter of *First about the Epidemic*, and *The First Question Is, What Are the Diseases Now Current;* the causes alone occupy *Is It from Divine Vengeance;* while the brief sets of pest rules emanating from the Strasburg physicians and from John Hake are concerned chiefly with prevention.

The first thing to do when the symptoms of plague appear is to have blood drawn by venesection or cupping. Ibn Khātimah's directions are to give immediately a drink of two ounces each of vinegar syrup and rose syrup mixed, and then have the bleeding take place where the pain is worst: if in the head, from the cephalic vein; if in the throat, from the basilic; if spread through the body, from the median. A good deal should be drawn, varying with age and strength, till the patient feels weak; but no effort should be made to continue it till the blood is of normal quality and tint, for in this disease the blood is thick and black, and its quality would not change if it were all drawn from the body. Often there mounts to the surface a thin green liquid or a gray mass, and this is a very bad sign. If the patient becomes faint before the required amount of blood is drawn, his face and extremities should be sprinkled with cold water until he revives, and then the bleeding should be finished. If he has not been in contact with the sick, his condition will probably improve and the fever fall, whereupon he should be given, carefully, according to the strength of the heart, apple syrup mixed with lemon syrup dissolved in rose-water and vinegar; afterward, broth prepared with peppermint; and then sour pomegranate. In case the sickness lasts or returns, and the patient has been in contact with the pest-smitten, there is small chance of curing him: the corruption is too far advanced and in most cases he dies; so the physician may as well give him something for his heart and leave him. When the disease is accompanied by blood-spitting there is no treatment, except possibly further bleeding; only one patient in Ibn Khātimah's experience has recovered. And for the illness in its other manifestations treatment has not much object, but since it is sometimes of avail, it will be set forth.

If bleeding has not removed the poison, the heart begins to separate the poisonous matter from the blood, and in the case of the bubonic form of the plague, to drive it to the emunctories. The patient ought to be bled moderately if the symptoms of the blood appear: these are blood vessels rich in blood, full pulse, temperature not too high, and something like choking attacks on the part of the sufferer. But bleeding is not advisable if there occur frequent bilious vomiting or agonizing diarrhea, and not if the sickness has lasted over two days, since it would then be too weakening, and hasten the end. If the symptoms call for bleeding, venesection should take place in the vessel indicated by the pain, from the opposite side if the pain is under the arm, the same side if it is behind the ear or in the inguinal region.

John of Penna requires bleeding and purgation as soon as the symptoms appear, laying more stress upon the latter. His only directions for bleeding are that it ought to take place from the side in which the pain is felt and not the opposite, since on account of the great malignity of the pestilence it is dangerous to draw it within; if the pain pass to the other side or is felt again in the first, bleeding should be repeated as often as the strength of the patient will permit. Gentile of Foligno's treatment, too, starts with bleeding: venesection of the median vein if the disease is taking no determined course; cupping in the places in which the open spaces are closed and if the imposthume appears below the open spaces. If the imposthume is in the neck or head, the cephalic vein should be cut in the two thumbs; if under the arms or in the right arm, the blood should be drawn from the pulmonary vein between the middle and ring fingers of the same arm; if in the heart, from the splenic vein in the right arm between the ring and the little fingers; if the left foot or the inguin, from the paralitic vein; if in the right foot, from the saphenic or vein of women. The bleeding should continue till the patient becomes faint, and should be repeated if the sickness remains lurking in the system, the patient being later strengthened with nourishing foods.

For the treatment of the buboes, Gentile recommends on the second day placing a large bleeding cup over the imposthume, and deep scarification with a knife; he inclines favorably also to cauterization. A strongly drawing plaster ought then to be placed upon the imposthume: an excellent one, which resolves the poison in a marvelous way, may be made from galbanum, the root of white lilies, human stercus, and *Consolida major*, ground between two tiles. Scabious is recommended for its efficacy against anthrax, and there are other recipes, chiefly of herbs, for plasters to draw out the poison, followed by several of value in drying out and cleansing the wound after the imposthume has been opened and the poison drawn out. Master Albert's contribution to the science of surgery is the suggestion that to the tumor of the glands be applied an old cock cut through the back;[50] and John of Penna, whose interest is evidently in the medical aspect of the subject, gives no directions for the opening and subsequent treatment of imposthumes.

Ibn Khātimah, however, shows himself something of an authority on the subject. The fourth day is the crisis in the bubonic form of the plague, and if it is passed safely, the matter is beginning to be separated from the heart and isolated, and the heart to be freed of it. This can be accomplished on the seventh day at the earliest, when one of several remedies for which he gives prescriptions should be applied to draw the buboes and hasten their ripening. When they are ripe and the blood in them is changed to

[50] Dr. A. C. Klebs has given me an interesting explanation of this treatment which, he says, was not uncommon in medieval medical directions. The cock was to be seized while crowing, the tip of the spine to be cut off, and the rear part of his body to be applied at once to the imposthume, simultaneously, if possible, with his being killed by suffocation.

pus, they should be opened with instruments as quickly as possible, for it is not dangerous, and after it is done safety against danger is reached. Prescriptions are given to heal the irritation; the worst thing that can happen is to delay the healing, for if the buboes are not resolved they may cause a recurrence of the disease. It is possible for them to be dissolved with the aid of medicaments, but operation is better, great care being taken, however, not to open them before they are ripe, a step which will result in the patient's sudden death, since the substance of the nodules is driven out of the heart and is in connection with it through the vessels. Consequently, should they be opened before their substance is isolated from the heart, localized in the organ, and changed to pus, the blood would be taken from the heart and its vital principle drawn into the connected blood-vessels. The result would be instant death. He cites here two instances, both related to him by others, of death's immediately following the untimely lancing of such nodules; and one case, told him by a trustworthy man who had heard it from Christian merchants traveling from Majorca to Almeria, of a physician who had dissected the imposthumes of a dead man, and had found that they were in connection with the heart through vessels filled with blood. This report Ibn Khātimah approves as correct and supported by investigation and theory. He concludes his directions for cure with instructions for treatment of black boils, which is much like that of buboes.

The remedies given are for constipation, to allay thirst, to stop diarrhea or vomiting—as digestive cordials—for faintness and weakness of the heart, for hoarseness in the larynx or heaviness in the throat. John of Penna suggests, to strengthen the heart, various electuaries, such as syrup of roses or juice of buglossa or borage, and sugar, with which may be mixed, for those especially requiring greater strengthening of the heart, precious stones, little corals, bones from the heart of a stag, and the like. Master Albert gives two recipes for internal use, one a syrup and one a confection, and for external application several odoriferous lotions or poultices for strengthening the heart: in all of these, herbal ingredients and precious stones and metals are about evenly divided. These two writers, like the others, advise foods easy of digestion, such as broths, the meat of pullets, partridges, and pheasants, and for drinks the juice of oranges or lemons, pomegranate wine, and other things tending to prevent corruption. Gentile of Foligno includes two potent digestive cordials, one a compound of pearls, emeralds, and other stones, to be ground with terra sigillata and mixed with rose-water in which has been rubbed a golden wand or golden ducats till the water has taken the hue of the gold; and the other, which seems even worse, potable gold. This is to be made of an ounce of best gold and eleven ounces of quicksilver, the two to be dissolved together over slow heat. The quicksilver is allowed to escape through the mouth of the alembic, after which the blackened gold in the bottom is mixed with forty-seven ounces of water of buglossa, stopped tightly in a glass alembic so that the air cannot get in or out, and kept for three days and nights continuously over an even fire. Then "if you know the art, you will find the water not consumed and the gold liquified, and this will be potable gold."

A welcome change from these mineral remedies are the herbal prescriptions of Ibn Khātimah. And especially pleasing is his injunction to avoid all strongly-working medicines that are related to the disease and that disturb nature, since they bring sickness and death before anything can be done to prevent it; and bad-tasting ones, since they are irritating to the constitution. Indeed, aside from powdered precious stones and potable gold, which could hardly have seriously affected many people, it is surprising how few prescriptions occur in the tractates which run decidedly counter to medical ideas of today. Bleeding was accepted, ingredients were prescribed which are no longer regarded as helpful, but the fantastic prescriptions and noisome mixtures which many casual writers seem to regard as inseparable from medieval medicine are notably absent. Whatever the remedies were, however, they were of slight avail against the disease, and the best advice was that with which Gentile concluded his chapter on surgical treatment: "Finally I conclude that to flee, as I have said, is best in this particular pestilence; for this illness is the most poisonous of poisons, and by its spread and blight it infects all."

Chapter IV

OTHER EFFECTS OF THE BLACK DEATH IN THE FIELDS OF MEDICINE, SURGERY, AND HYGIENE

In addition to the plague tractates evoked by the Black Death from medical men of learning, there were other consequences of the great mortality for them and their profession. First we may inquire as to the number of physicians and surgeons who were among its victims—a question presenting peculiar difficulties because of the scarcity and irregularity of records kept while the epidemic was at its height. The chronicles usually note deaths of political and ecclesiastical authorities, but are inclined to refer in general terms, or frequently not at all, to the ravages of the pestilence in other professions. So it is only from chance comments here and there, from such town, church, and university records as we have, and from the disappearance at about that period of men who had been active till then that we can draw any conclusions. The attempt to obtain any sort of an estimate of the mortality in the medical profession must be based upon a cross-sectional study of limited groups, which may possibly be considered illustrative of the whole.

Of the fifteen named authors of the plague tractates which we have discussed as written between 1348 and 1350, Gentile of Foligno died in 1348 of the pestilence, and John Hake of Göttingen, in 1349 with no cause of death assigned.

Nine[1] lived through the pestilence, and four[2] were not heard of afterward; as they are equally unknown to us before writing their treatises, there is no reason to attribute their disappearance to the epidemic.

Statistics in regard to the medical staff of the pope at Avignon at this time are somewhat more definite. Clement VI had during his pontificate (1342–1352) nine physicians and three surgeons. Of the physicians, three were last heard of in 1348: Albert of Würzburg made his will 10 March, 1348; John of Florence entered his marginal signature upon receipt of his last salary 26 July, 1348, and in 1353 his library had been incorporated with that of the pope; in 1348 William of Lavetajo performed his only recorded act when he drew up a prescription for pills against the pestilence. The same was true of two, or we might say both, of the surgeons, since the third, John of Parma, was engaged to succeed Peter Augerii, who died in 1348; the other, John of Genoa,[3] was paid his last salary 31 May, 1348. So that the presumptive evidence here is that from among the pope's medical advisers five died out of twelve.[4]

[1] John of Penna, Dionysius Colle, Simon of Covino, Ibn Khãtimah, Ibn al-Katīb, and four of the authors of the Strasburg *Treasure of Wisdom.*

[2] James of Agramont, Master Albert, Alfonso of Cordova, Henry of Lübeck.

[3] He is generally thought to have been the same John of Genoa who compiled *Canones eclipsium* in 1332, and wrote *Investigatio eclipsis solis anno Christi 1337*. P. Duhem, *Le Système du monde,* Paris, 1913–1916, Vol. 4, pp. 74–75.

[4] P. Pansier, "Les médecins des papes d'Avignon," *Janus,* XIV, 405–434, makes a study of the physicians and surgeons of the popes from 1308 to 1403, based on Marini's *Degli archiatri pontifici* (1784), and supplemented by work in the records of Avignon. The physicians given by Marini whom he accepts are Stephen Seguini, John of Florence, Stephen Ancelini (last date 1345), Raymond Rainaldi of Vinario, William of Lavetajo, Laurence of Biaze (d. 1369), John of Marescalia, Guy of Chauliac, Albert of Würzburg; the surgeons

In Ibn al-Khatīb's list of fourteenth-century men of learning in the kingdom of Granada, there are included ten physicians who died after 1347 and a monk, adept in chemistry, who seems to have exercised the functions of a physician. One of them, Alschaphra, especially skilled in botany, is without date; four, all of whom survived the Black Death, have already been treated;[5] one died in 1352, and two in the sixties; the monk and chemist, Alansareus, commonly called Alsanna, in 1348; in 1349 two, Alkhagragi, "highly skilled in medicine, using as easy and mild medicaments as possible, rarely resorting to the cruel instrument," for whose death no cause is given, and Alsekuni, commonly called Ebn Allulu, who died of the Black Death. In the time when the pestilence was abroad in Granada, then, we have three physicians dying out of ten for whom dates of death are given.[6]

Rolls and records of universities are of small help in the matter.[7] There are none from the English universities that are of assistance; none from Bologna in the fourteenth century before 1384,[8] except the jurists' statutes of 1317–1347;[9]

---

are Peter Augerii, John of Genoa, and John of Parma. He is inclined to reject two physicians given by Marini who are not included here: James Capelluti and John of Alesto.

[5] See above, pp. 18–21; 26–28.

[6] Published by M. Casiri, *Bibliotheca arabico-hispana escurialensis*, Madrid, 1770, Vol. 2, pp. 73–111. Part of the MS was lost, so that the list begins with the name Mohamad. See above, p. 19. Casiri's spelling is kept for the Arabic names in this paragraph, except Ibn al-Khatīb.

[7] For a study of the effect of the Black Death on education and universities, see below, Chapter Six.

[8] C. Malagola, *Statuti delle università e dei collegi dello studio bolognese*, Bologna, 1888, p. xviii.

[9] H. Denifle, "Die Statuten der Juristen-Universität Bologna vom J. 1317–1347," in *Archiv für Litteratur und Kirchengeschichte des Mittelalters*, III (Berlin, 1887), 254–393.

none from Montpellier before 1362, and though one of that date is the roll of the faculty of medicine, it bears only twenty-eight names, chiefly of bachelors in medicine;[10] from the University of Toulouse there is a roll, also of 1362, in which appears the name of one master in medicine.[11] It is not known whether Perugia had a faculty of medicine before the middle of the century,[12] and in such rolls as exist of the smaller universities the faculties are not given separately.

From Padua we have relatively full lists of names for the last half of the fourteenth century. Before 1347 there are but three professors of medicine, surgery, and astrology given by Gloria; from 1347 to 1349 there were also three, if we accept Nicholas Santa Sophia, whom Gloria considers doubtful. One of the two he admits was James Dondi, the other a Master Robert, who was lecturing on medicine and philosophy at Padua in 1348, when Venice called him to her service.[13] For a year or so there must have been a lack of professors of medicine, surgery, and astrology at the University of Padua, with James Dondi holding the position alone, till in 1351 the names of five new professors appear. Others were added, with the result that a decade after the Black Death the number was twelve.[14]

There is, however, no indication of deaths in the medical

[10] *Cartulaire de l'Université de Montpellier,* Vol. I, Montpellier, 1890, No. 120.

[11] M. Fournier, *Les Statuts et privilèges des universités françaises depuis leur fondation jusqu'à 1789,* Vol. 1, Paris, 1890, No. 656.

[12] P. D. V. Bini, *Memorie istoriche della Perugina Università degli studj e dei suoi professori,* Perugia, 1816, p. 173.

[13] A. Gloria, *Monumenti della università di Padova,* Padua, 1888, Vol. 1, pp. 369–371.

[14] A. Gloria, *Monum. Uuiv. Pad.,* I, 369–390.

faculty of Padua, nor at the University of Paris. A roll of the four faculties of the latter university, dated 19 May, 1349, Avignon, but taken in accordance with resolutions passed 6, 7, 8 May, 1348, Paris, shows the faculty of medicine to consist of forty-six members;[15] a roll of 1362 gives twenty-six,[16] and one of 1387, twenty members.[17] Obviously the question here would be whether the first roll was taken immediately after the vote or later, when the epidemic had reached Paris: the messengers bearing it to the pope reached the curia 24 November, 1348,[18] and we know that the plague, which had begun in Paris in the summer of 1348, raged through the fall and into the early winter. But in a study by Wickersheimer of the physicians of the English nation at the University of Paris there is nothing to indicate that pestilence occurred in the fourteenth century: five medical students determined, were licensed, became proctors of the nation, or received benefices in 1349, their activities having gone on apparently without interruption for several years previously.[19] The same author, in a search through the records of the University of Paris for enactments concerning epidemic diseases, finds the first mention of epidemic in the *Commentaries* of the university occurring 8 November, 1399, when the university was suspended on account of the "mortality."[20]

[15] Denifle and Chatelain, *Chartularium Universitatis Parisiensis,* Vol. 2, Nos 1152, 1164.

[16] *Chart. Univ. Par.,* Vol. 3, No. 1264.

[17] *Chart. Univ. Par.,* Vol. 3, No. 1540.

[18] *Chart.,* Vol. 2, No. 1152.

[19] Wickersheimer, E., *Les Médecins de la nation anglaise de l' Université de Paris aux XIV et XV siècles.* Paris, 1913.

[20] E. Wickersheimer, "Les Maladies épidémiques ou contagieuses (peste, lèpre, syphilis) et la Faculté de Médecine de Paris," in *Bull. soc. franç. hist. méd.,* XIII (1914), 21–30.

Documents of another kind seem to tell a different story. Simon of Covino, in his poem about the pestilence, composed in 1350 in Montpellier and Paris, says that at Montpellier, where there was a greater number of physicians than elsewhere, scarcely one escaped.[21] Gabriel de Musis, a notary of Piacenza who died in 1356, tells us in his contemporary *History of the Sickness or Mortality Which Was in 1348*[22] that in Venice twenty out of twenty-four very excellent physicians died in a short time. Cecchetti quotes from the archives of the same city for the early months of 1348 that the ranks of the doctors there were so thinned "that it can be said that not one, as it were, remained."[23] Flight and fear as well as death seem to have played a part in stripping the city of its medical talent, for it is remarked that she had to seek more valiant men to take their places, and that some shut themselves in their houses. This is the time, 22 May, 1348, when Master Robert, Paduan professor of medicine and philosophy, was procured as physician by Venice, and assigned the salary of 150 lire formerly paid Master Amedeus, who was now dead.[24] The following year he, together with a Master Berardo, son of Conegliano, and Master Bressanino, son of the late Bettino, were chosen at Treviso, "in the lack of all physicians," with the annual salary of 728 lire. The predecessor in office of

[21] See above, p. 31.

[22] Published by Henschel in Haeser's *Archiv für gesammte Medizin*, Jena, Vol. 2, pp. 25–59; Tononi, in *Giornale ligustico*, Genoa, Vol. IX, p. 139 ff.; and in part by H. Haeser, in the appendix of his *Lehrbuch der Geschichte der Medizin und der epidemischen Krankheiten*, 3d edition, Jena, 1882.

[23] B. Cecchetti, "Med. in Ven. nel 1300," *Archivio Veneto*, XXV (1883), 377. The rest of the data in this paragraph, except those for which other references are given, are taken from this work, pp. 361–381.

[24] Gloria, *Mon. Univ. Pad.*, Vol. 2, p. 28.

Bressanino had been Master Federico, who had obtained his doctorate at Padua, and had practiced medicine there for a time with Professor Mondino of Forlì. In July, 1345, he was taken into the services of the commune of Treviso with the yearly stipend of 150 lire. The fact that four years later, after his death, his successor received a salary nearly five times as large as his own had been is striking testimony to the sudden and urgent need of skilled physicians.[25] The death of three other Venetian physicians is indicated by a resolution of the senate, 14 December, 1348, to furnish the city "with good physicians, such as had been Master Nicolino, Master Pagano, and Master Peter of Venice, from among the more notable and the better established." On 12 January, 1349, Marco Leone, "a famous doctor," at that time in Perugia but originally from Venice, offered to return to his native city on account of the scarcity of physicians there.

From the *Memoriali* of the Notarial Archives of Modena, Signor Bertoni, assisted by Signor Vicini, has published the items referring to the fourteenth-century physicians and surgeons of Modena.[26] A study of those from the decade before the Black Death through the third quarter of the century reveals several points of interest. The majority

[25] Gloria, *Mon. Univ. Pad.*, Vol. 1, p. 448. Gloria makes the year 1348 instead of 1349 as given by Cecchetti, but the latter seems correct, since it was in 1348 that Master Robert was chosen physician at Venice at the stipend of £150, and he was one of the three selected by Treviso after the death of Federico, at the greatly increased salary.

[26] G. Bertoni, and E. P. Vicini, "Medici modenese del secolo XIV," pp. 133–141 of *Atti e memorie della Reale Deputazione di storia patria per le provincie modenesi.* Ser. V, Vol. 4 (1905). The list of physicians and documents concerning them is an appendix to an article by Bertoni in the same periodical, pp. 125–132, "I codici di un medico modenese del secolo XIV (Jacopino Cagnoli)."

of the men belong to three leading professional families of Modena: the Candelini, the Cesi, and the Actolini; the two first figuring more prominently from 1337 to 1349, the third, in the second half of the century. Another noticeable feature is that, though five of the eleven are mentioned more than once, not a single name which appears before 1350 occurs after that date. In the decade before the outbreak of the pestilence there are seven physicians and surgeons; during the twenty-five years following, five.

Henry Candelini, surgeon, is officially mentioned only in 1337, but he is referred to as deceased in a later passage which has to do with his son Peter. Peter also was a surgeon, and appears three times in the archives, in 1339, 1343, and for the last time when he made his will in 1348. Henry had two other sons, both of whom were physicians: Cherubinus Candelini, whose name occus in 1337 and 1338; and Marcellus, for whom we have no dates, since he is known only through references to him in passages concerning his son James, a distinguished doctor of arts, medicine, and physics, in the last quarter of the century.

Peter Cesi (or Zesi), physician and professor, is mentioned in 1328 and 1337. He was son of the deceased jurist Albert Cesi, and his sepulchral inscription seems to show that he died in 1349.[27] The will of another Cesi, Nicholas,

[27] The inscription runs:

"Hunc artes, hunc Sancta Dei medicina coronat
Nomine mirifico gelidum quem clausa supinant
Saxa virum, toto latio qui cognitus orbe
Scaligeros curabat heros, cui nomine Petrus
De Cesis, Doctorque probus hic longe quiescens
Scansit ad almiacos felix sine labe triumphos
Mille Dei centumque dabant ter tempora Cristi
Ante novem deciesque quater tunc Madius annos."

—Bertoni, "Medici modenesi," p. 136. See also G. Tiraboschi, *Biblioteca modenese,* Vol. 1, p. 24, and, by the same

is given in 1349, but there is no indication of his profession. He was father of Geminian Cesi, called in the records of the latter part of the century a famous doctor and professor of arts and medicine. The other two physicians of the earlier period were brothers, Gerard and Francis Caxano, sons of the late Eliga Caxano, who are named together in an item of 1339. Pax Zandori, physician and professor, son of the late John Zandori, and John Gratie, surgeon, are mentioned in 1350; the name of the former appears on the roll of the University of Paris, faculty of medicine, of 1349, asking for a benefice in a church of Ferrara. Francis Actolini is to be found in documents of 1350, 1363, and 1381, in which he is said to be a distinguished professor and doctor of arts and medicine, son of the late Actolini, himself the son of James Actolini. Francis' son, Actus Actolini, followed in his father's footsteps, figuring as physician and professor in the records of his native city during the last quarter of the century. John Novellus, a surgeon, appears once, in 1362, and Peter della Rocca[28] twice in the same year, 1362, in connection with his will and codicils. While these notarial archives do not, in their nature, record causes or even dates of death, we may take it that at least two, Peter Candelini and Peter Cesi, of the six medical men who are mentioned for the last time during the decade ending with the Black Death, died of the pestilence.

The difficulty in determining the cause of death of individual physicians may be illustrated by the two cases of Freidank and Bertrucci. Freidank was physician of Günther of Schwarzburg who was offered the German

author, *Dizionario topografico-storico degli stati estensi,* Vol. 1, p. 203.

[28] *Chartularium Universitatis Parisiensis,* Vol. 2, No. 1164.

throne at the death of Louis IV in 1347, and was elected king by four of the imperial electors, who were opposed to Charles of Bohemia. Günther was defeated by Charles at Eltville, sold his claims to him, and died soon afterward.[29] The current story, repeated as a historical fact, was that Freidank had been persuaded to poison Günther, and had himself died as a result of the king's forcing him also to drink the poison.[30] Dr. Römer-Büchner announced in 1857 that he had found, by research in Frankfort, that both Günther and Freidank had died of the Black Death, Freidank 14 April, 1349, and Günther 18 June of the same year.[31]

Bertrucci, Vertuzzo, Bertruccius, or Nicolaus Bertucci, as he was variously called, was a skilled anatomist, pro-

[29] These facts are given in whole or in part by several contemporary German writers. The most complete account I have found is contained in the contemporary annotated history of Michael de Leone of Würzburg, published by J. F. Böhmer, *Fontes rerum germanicarum,* I, 477–478.

[30] It is thus that the story of the poisoning, including Freidank's name, is given in the chronicle of Matthew Nuewenburg, 1273–1350 (with continuation till 1378), in Böhmer, *Fontes*, IV, 269. A brief reference to the poisoning occurs also in the anonymous *Chronicon de ducibus Bavariae,* 1311–1372, written in 1372: While Günther was valiantly striving against Charles "nec ab eo ullo modo flecti posset mortuus est de toxicatione, quam predicto rege Bohemiae ut multi dicunt procurante, quidam sibi physicus propinavit." Böhmer, *Fontes,* I, 145.

[31] Römer-Büchner, *Allgemeine Zeitung,* 1856, No. 52. See also Stricker, *Geschichte der Heilkunde in Frankfurt am Main,* Frankfurt am Main, 1847. Albert Moll, "Der schwarze Tod in Würtemberg," in *Medicinisches Correspondenz-Blatt des Würtembergischen ärtzlichen Vereins,* Band 27, No. 32, 5 Oct., 1857; H. Haeser, *Lehrbuch,* p. 123. I am unable to give Dr. Römer-Büchner's proofs, since I have not succeeded in finding his work either in this country or in the British Museum in London. My information is obtained from the article of Dr. Moll, who accepts his conclusions, as does also Haeser.

fessor at Bologna, where Guy of Chauliac studied under him, and author of medical works of which three have been printed.[32] His death in the latter part of 1347 has been attributed to the Black Death,[33] but the account in the *Chronicle of Bologna,* which was written by a contemporary Friar Minor, Bartholomew della Pugliola, points to a slightly earlier "great plague" as the cause. In the latter half of 1347, he writes, there was a great plague and famine, from which many people died. He names eleven distinguished victims, among them the eminent physician Master Vertuzzo, the jurist Master James Butrigari, and Masters Azzo Contughi and Peter Buonpieri, doctors.[34] Under the year 1348 he tells of the two Genoese galleys which brought the pestilence from Constantinople and Pera to Sicily and Messina, and of its spread into Italy. "In the month of March began a great plague in Bologna, and lasted till St. Michael in September. It was so fierce that it was estimated that of five three and more died, and it was greater than can possibly be told. Through the said pestilence died the most famous doctors of Bologna, such as Master John Andrea, and others." And with that one name, as if unequal to embarking upon the list of fatalities, he stops.[35]

There are several physicians besides Freidank who were connected with royal households and were last heard of at the time of the plague. Guy Vigevano of Pavia was physician of Emperor Henry VII, later of Jeanne of Burgundy, wife of Philip VI of France, and still later of both Philip and Jeanne. In 1335 he wrote a *Thesaurus of the King of*

[32] Tiraboschi, *Stor. d. lett. ital.,* V, 401; Pagel *Handbuch,* I, 671.
[33] G. T. Fort, *Medical Economy during the Middle Ages,* New York, 1883, p. 435.
[34] "Chronicle of Bologna," in Muratori, *Scriptores,* Vol. 18, Col. 402.
[35] "Chronicle of Bologna," Muratori, Vol. 18, col. 409.

France of the acquisition of the holy land, also of the health of his body and the prolongation of his life, and also with advice for guarding against poison. The first part is a "Book on the conservation of the health of an old man," and the second an illustrated treatise on the military art. In the introduction to the *Thesaurus,* Guy is called physician of Queen Jeanne; but the king evidently appreciated the efforts put forth in his behalf in that production, for the author is said to be physician to both Philip and Jeanne in a work composed in 1345, a collection of medical treatises of which one is his *Anatomy,* illustrated with eighteen figures in color.[36] The last notation of payment made him, in the treasury records of Philip VI, is July 1349,[37] so he probably died that summer of the pestilence, a fate which overtook Queen Jeanne the following December. James Albini of Moncalieri, physician of James, prince of Achaia,[38] wrote a work, *The Maintenance of Health,* which also includes the hygiene of travel and armies. The last date we have in his life is 1349. Geoffrey Isnardi, physician of Pope John XXII (1316–1334) and titular physician of Benedict XII (1334–1342), died at Avignon 26 July, 1348,[39] and to the north, in Upper Auvergne, Gregory Talayzat, physician at Saint-Flour, dis-

[36] E. Wickersheimer, "L'Anatomie de Guido de Vigevano," *Archiv,* VII, 1–25.

[37] J. Viard, "Journaux du trésor de Philippe VI de Valois," in *Documents inédits de l'histoire de France,* 1899. Guy was paid as physician of the queen in both June and July, 1349.

[38] G. Carbonelli has written a work which I have not been able to obtain, *Il "De sanitatis custodia" di Maestro Giacomo Albini da Moncalieri, con altri documenti sulla storia della medicina negli stati sabaudi nei secoli XIV e XV.* Pinerolo, 1906. It is fully reviewed, however, by M. del Gaizo, in *Janus,* XII, 415–417.

[39] P. Pansier, in *Janus,* XIV, 405–406.

appeared in the midst of the great pestilence of 1348.[40] In Bologna Albert Zancari, professor of medicine at the university from 1326 to 1347,[41] is not heard of after that year, though it would seem that a special tribute was paid his memory in a statute of the commune of Bologna in 1352. By this it was arranged that the town would provide fixed salaries, in addition to what was regularly paid by the students, for two citizen doctors to give morning lectures on medicine for a year, with the stipulation that one of the two was always to be, as long as it pleased him, Master Fabian, son of Master Albert Zancari.[42]

In wondering who the numerous physicians in Montpellier who died of the pestilence may have been, we find the names of certain practitioners of that place who were active till some ten or twelve years before the pestilence, but of whose deaths we know nothing. Such were Bonetus Mote or Lanfranci, son of Lanfranc[43] the surgeon, who was physician of Benedict XII and was cited by Guy of Chauliac as a renowned surgeon;[44] Raymond of Molière, chancellor of Montpellier in 1338, mentioned by Guy of Chauliac, and author of a work, which has been published,

[40] M. Boudet, and R. Grand, *Étude sur les épidémies de peste en Haute-Auvergne, XIV–XVIII siècles*, Paris, 1902, p. 38.

[41] Tiraboschi, *Stor. d. lett. ital.*, V, p. 402.

[42] C. Malagola, *Statuti delle università e dei collegi dello Studio Bolognese*, Bologna, 1888, p. 154, note.

[43] Author of *Chirurgia magna* and *Chirurgia parva*, and master of John Yperman.

[44] Pansier, in *Janus*, IX, 542, and XIV, 413; E. C. Van Leersum, in *Janus*, XVIII, 207. But in the roll of the medical faculty of the University of Paris, taken at some time between May and November, 1349, and sent to Pope Clement VI, I find the name Bonito Moteti of the diocese of Clermont on the Meuse. This seems to indicate that if Bonetus Mote died in the forties, it was during the epidemic of 1349 in Paris. *Chart. Univ. Paris.*, Vol. 2, No. 1164.

on the causes of sterility;[45] Stephen Arnaldi or Arlandi, cited by Guy of Chauliac and Valescus of Taranta, author of medical and surgical works, some published and some in manuscript, vice-chancellor of Montpellier in 1319, and flourishing till at least 1340;[46] and Jordan of Tours, professor at Montpellier at various times between 1313 and 1335, mentioned by Guy of Chauliac and Valescus of Taranta, and author of two works which still exist in manuscript.[47]

To the south, also in Aragon, John Amelio was flourishing in 1336 as chief personal physician of Alfonso IV, and in 1343 Peter Ros of Ursins was given a donation by Peter IV of Aragon.[48] In the same year, 1343, Thomas Scelling, a surgeon, probably of Flanders, finished writing his *Surgery*, and disappears from our view.[49]

Some medical practitioners died toward the end of the Black Death period, but apparently not of the plague: Nicholas Santa Sophia died in 1350;[50] about the same year died Nicholas of Reggio, who was the first to translate Greek medical works directly into Latin without Muslim influence;[51] in 1350 died also William Grisaunt or English,

[45] Pagel, in *Janus*, VIII, 531–537; Pansier, in *Janus*, X, 8.

[46] Pansier, in *Janus*, X, 9–10.

[47] Pansier, in *Janus*, X, 3.

[48] Both of them are given by L. Comenge, in *Janus*, VIII, 326–328.

[49] E. C. Van Leersum, "Master Jan Yperman's Cyrurgia," in *Janus*, XVIII, 198. According to Leersum's data, John Yperman seems to have died so early in the fourteenth century that he antedates our period.

[50] Physician of Padua, founder of the distinguished medical family of Santa Sophia.

[51] G. Sarton, *Introduction to the History of Science*, Vol. 3. This volume is still in manuscript. Tiraboschi, *Stor. d. lett. ital.*, V, 422, says that Nicholas' works were praised by Guy of Chauliac as seeming better than those translated from the Arabic; he had, Guy said, "sent them to us in the Curia."

an English physician of Merton College, Oxford, who studied at Montpellier rather late in life, and afterward practiced in Marseilles;[52] and Peter of Solo or of Arles, a surgeon and physician mentioned by Guy of Chauliac, died before 1352.[53]

Of all these possibilities, we can reasonably assume the death of only eleven as a result of the Black Death,[54] and most of these are based on date of death without definite statement of cause. Twelve of the physicians and surgeons who wrote on the pestilence in the second half of the century, besides those whose names have already been used in computations of mortality, were practicing at the time of the Black Death.[55] The same is true of twenty-four from whom we have no writings on the pestilence,[56] so that of fifty-one men who are known to have been prac-

[52] *Dictionary of National Biography*, VIII, article on William Grisaunt; Pansier, in *Janus*, IX, 594; J. Astruc, *Histoire de la faculté de médecine de Montpellier*, Montpellier, 1862, p. 184; F. A. Gasquet, *The Great Pestilence*, London, 1893, p. 35.

[53] Pansier, in *Janus*, XIV, p. 427.

[54] The four Venetian physicians, Amedeus, Nicolino, Pagano, and Peter; Albert Zancari; Federico at Trevisa; Freidank; Geoffrey Isnardi; Gregory Talayzat; Guy Vigevano; James Albini.

[55] Francis Casini of Siena; Gallus of Strahov; John Arderne; John of Burgundy; John Dondi; John of Glogau; John Jacobi; John Santa Sophia; John Tornamira; Maino Maineri; Thomas of Breslau; Thomas Garbo.

[56] Alexander of Rocafort; Anthony of Curtarola; Anthony Lio of Padua; Anthony of Pernumia; Barnabas of Reggio; Bartholomew Campo of Padua; Bernard Alberti; Enrigetto Lio of Padua; Francis, son of Peter Zannelli, of Bologna; Francis, son of Philip, of Foligno; Gandolfus of Cremona; Gerald of Solo (or Bierne); Guy Bagnolo of Reggio; Henry of Regensburg; Hugo (Ugoccione) of Rio; James of Arquà; James Dondi; John of Gaddesden; Moses ben Joshua of Narbonne; Mundinus of Forli; Peter Pestagalli; Peter of Reggio; Raymond Salayronis; William Ghezzi.

ticing physicians when the epidemic occurred,[67] eleven died when it was prevalent in their communities, four soon after it had passed and before it had entirely ceased, and thirty-six survived.

It is gratifying to one who has acquired respect and liking for these earnest, and often gifted, scholars of a past age to note that while, as far as we know, only two of the writers of plague tractates died, two—Dionysius Colle and Guy of Chauliac—tell us that they contracted the plague in 1348 and, though their recovery was despaired of, were cured by their own treatment. In view of the deadliness of the disease, not only during that fearful outbreak but even today if the Yersin-Roux serum is not used in time, this record would seem to be a feather in the cap of fourteenth-century medicine.

A noticeable point about the lives of many of the physicians is the number of instances in which the Black Death seems to have formed a turning point, or at least to have played an important part, in their careers. Ibn al-Khatīb, on the death by pestilence in 1348 of Ibn al Gaijâb, wezîr of Sultan Jusuf of Granada, succeeded him in that office and remained wezîr till a few years before his death in 1374. John Dondi became personal physician of Charles IV in 1349, and in the same year Gallus of Strahov appears to have found a place at the court of the same monarch. In 1348 John of Tornamira began his practice at Montpellier, where he remained till 1369; and John Arderne states that from the year of the first pestilence, 1349, till 1370 he lived at Newark, in the county of Nottingham. Guy of Chauliac was papal physician at Avignon from 1348

[67] Since this excludes the physicians of the period (38 in number), who have been named before in estimating mortality, the total becomes 89.

till his death in 1368, and the debut of another surgeon, John of Parma, as a member of the papal staff in 1348 was due to the death by pestilence of a papal surgeon, Peter Augerii.

The fact that the last three of these men were surgeons (two of them, Guy of Chauliac and John Arderne, very distinguished ones), brings to attention a momentous change that occurred in the field of medicine in the fourteenth century, and which may have been accelerated by the Black Death. Till about the end of the thirteenth and the beginning of the fourteenth centuries, surgery had never been held in esteem, nor, except for a period of brilliance in Roman times, had it developed beyond a rudimentary stage. As intellectual activity and curiosity increased in Europe, and the dependence of accurate knowledge upon experimentation became more evident, enterprising spirits began to break through tradition and prejudice. In the thirteenth century the practice of dissection was begun by Mundinus of Bologna;[58] he wrote an *Anatomy* which was the first work of the medieval period devoted entirely to the subject. His pupil and successor was Bertruccius, who, in turn, was instructor of Guy of Chauliac. Others who blazed the trail were John Gaddesden, Lanfranc, Henry of Mondeville, and John Yperman, the last three authors of works on surgery in the early fourteenth century.

But the *Great Surgery* of Guy of Chauliac and the *Practice* of John Arderne show a decided advance over the earlier works of the century, and from the middle of the

[58] *Histoire littéraire de la France,* XXVIII, 333. Mundinus of Bologna, who died about 1325, should not be confused with Mundinus (or, in either case, Mondinus) of Forli, professor at Padua later in the century. See also Tiraboschi, *Stor. d. lett. ital.,* V, 414–422.

century a double tendency is increasingly discernible: to break down the barriers between physicians and surgeons, and to standardize the profession of surgery. For instance, in Venice it was against the law for surgeons to practice medicine, and 31 December, 1348, a fine was inflicted upon Andreas of Padua because, in the mortality of 1348, he went out as a physician and cured over a hundred of those smitten with the pestilence; this the Venetian judges attributed to accident rather than to wisdom.[59] Yet in September, 1349, Nicholas of Ferrara was invited to Venice from Padua and praised for his knowledge and experience in surgery: both of the cities in which he had practiced, it is said, had profited by his knowledge of diseases of the lower parts of the body [*crepatorum*], which required a specialist.[60] Guy of Chauliac, though really a surgeon, was physician of Clement VI; and John of Parma, starting in 1348 as papal surgeon, later became physician and surgeon. However the same thing is true of his predecessor, who in 1348 was called physician.

In England an indication of the trend toward raising surgical standards was the formation of surgeons' gilds. The Oxford gild was incorporated by Dr. Northwood, chancellor of the university, in 1348; it included barbers and waferers, but lectures in surgery began to be given about 1350. In 1362 it was decreed that every London surgeon must belong to a gild; further regulations were enacted twice later in the century, and before its close gilds of surgeons had been founded in half a dozen other towns.[61]

[59] B. Cecchetti, "Medicina in Venezia nel 1300," *Archivio veneto*, XXV, p. 369.

[60] Cecchetti, in *Arch. ven.*, XXV, p. 373.

[61] G. Parker, *The Early History of Surgery in Great Britain*, 1920, pp. 56–63.

The first judicial post-mortem was held in 1302 at the University of Bologna, where Mundinus had earlier inaugurated the practice of anatomical dissection; post-mortems occurred also in other universities, Gentile of Foligno holding one publicly in 1341 at Padua. But the first recorded instances of public officials, other than educational authorities, decreeing autopsies seem to occur during the Black Death. In a letter written from Avignon by a cleric of the Netherlands, the statement is made that "anatomical dissections have been performed by physicians in many states of Italy, and also in Avignon, by order and command of the pope, that the origin of this disease might be known, and many bodies of the dead have been opened and dissected."[62] One city in Italy which took such a step was Florence, for in a list of public expenditures dated 30 June, 1348, there is an item "for giving more corpses to physicians who requested them, in order to be able to learn more clearly the diseases of the bodies."[63] Public dissections were decreed by several other cities during the century.[64] In 1376 the duke of Anjou, governor of Languedoc, issued an ordinance to the officers of justice in that province: they were to suppress in their districts the illegal practice of medicine, and to send every year to the medical faculty of Montpellier for dissection the corpse of an executed criminal. The purpose of both decrees is set forth: it is an attempt to save what mortality, epidemics, and war have left of the population; dissection

[62] "Breve chronicon clerici anonymi," in J. J. De Smet, *Récueil des chroniques de Flandres,* III, 14–18. The writer goes on to say that it has been found that all who die so suddenly have an infected lung and spit blood.

[63] F. Carabellese, *La Peste del 1348 . . . in Toscana,* Florence, 1897, p. 48, quoting from Gherardi, *Misc. fior.*, p. 158.

[64] F. Baker, *Johns Hopkins Hospital Bulletin,* Vol. 20, Baltimore, 1909, p. 331, note.

is desirable, since experience is mistress of affairs, and darts foreseen usually cause less damage, and visible dangers can more easily be avoided, than the hidden and the unknown.[65] The last part of the fourteenth and the early fifteenth centuries were a period of renown for the medical faculty of Montpellier, for it then combined the knowledge of surgery with that of medicine. But Pansier calls attention to the fact that after this it again neglected the former science, with the result that it rapidly lost the brilliance which had resulted from the union of the two.[66]

Public sanitation is connected, at least indirectly, with medical affairs, and was decidedly affected by the Black Death. Public measures to combat the plague began in Italian cities in 1348, and developed by the seventies into the quarantine, which dates from this time, and by 1402 into the establishment of a lazaretto by Venice. This city, on a Sunday in March, 1348, elected a commission of three to suggest plans for the public safety. On the following Thursday they made recommendations to the Great Council, at a meeting open to all officials and judges, for measures to prevent the corruption of the city, dealing chiefly with the disposal of bodies.[67] The problem became so acute that the city purchased ships at a high price and carried the dead to islands.[68] In Florence, where the ravages of the plague were so terrible that it is sometimes called "the Florentine pestilence," the *gonfaloniere de giustizia* and the twelve *buoni uomini* chose eight of the wisest and most respected citizens, whose names are pre-

[65] M. Fournier, *Les statuts et privilèges des universités françaises*, Paris, 1890–1894, Vol. 2, No. 1020.

[66] Pansier, in *Janus*, X, 58–59.

[67] Cecchetti, in *Arch. ven.*, XXV, 377–378, from the city archives.

[68] "Historia Cortusiorum" (Contemporary), in Muratori, *Scriptores*, Vol. 12, col. 926–927.

served in the city records, to exercise a sort of dictatorship. This was the first office of public health, Carabellese comments, in the history of the republic; its duties were to see to the removal of decaying matter and infected persons from the city, and to supervise the markets.[69] The priors[70] of Florence at this time took into their service physicians to study the nature of the disease and suggest to them remedies to adopt.[71] Steps for protection against the epidemic suggested to the authorities of Perugia by Gentile of Foligno and the college of physicians there have already been discussed.[72] Ragusa put the matter into the hands of her count, decreeing that he was to see to medicines, physicians, and other preventive measures.[73]

Embargoes of various kinds, to keep out of the city infected persons or goods, were set up in 1348. As early as 13 January of that year Lucca issued a decree that no Genoese or Catalan, or anyone who had within the past year been in any city or parts of Romagna, should dare enter the city of Lucca, under penalty of the property and person of him who should violate the decree.[74] The Great Council of Venice forbade the bringing of a sick foreigner into the city on pain of imprisonment, burning the ship, or other penalty,[75] and as the pestilence advanced cities prohibited entry to all from without, with the result

[69] Carabellese, *La peste del 1348*, pp. 44–45, from the Florentine city records. He notes that there had previously been an official to expel lepers from the city.

[70] The priors were secular authorities chosen for the first time in 1282, and representing the gilds in the government of the city.

[71] Carabellese, p. 58. He does not give his authority.

[72] See above, p. 12.

[73] K. Lechner, *Das grosse Sterben in Deutschland*, Innsbruck, 1884, pp. 65–69, quoting from *Monum. spect. hist. Slav. merid.*, XIII, 11.

[74] S. Bongi, *Bandi lucchesi del secolo XIV*, Bologna, 1863, Vol. 1, p. 194.

[75] Cecchetti, in *Arch. ven.*, XXV, 377–378.

that merchants could not travel from place to place.[76] Matters finally reached such a point, Guy of Chauliac tells us, that guards were kept in the cities and villages to prevent the entry of anyone who was not well known.[77] A document of Pavia, dated 1387, is preserved, providing for the payment of seven officials who had been deputed to stand at the city gates for that purpose.[78]

Occasional glimpses afforded of the feeling between neighboring cities or regions make such protective measures seem natural. The Genoese galleys brought the pestilence to Messina in October, 1347, and so great were its ravages that the inhabitants left their town, and, other places refusing to receive them, were forced to camp out among weeds and vines. They sought especially Catania, hoping for assistance from the Blessed Virgin Agatha there, but the population of that town not only refused to admit them, but tried vainly to keep the patriarch from bearing some relics of Agatha to Messina. Next the panic-stricken people went to the Blessed Virgin of Scala, six miles distant, and chose a priest to bring her to their city, but she, having been brought unwillingly, would not help them. They scattered over Sicily and into Calabria, but all refused even to speak to them, so that "not to speak to anyone as if he were from Messina" became a proverb.[79] During the same pestilence it is recorded that the mortality was so great among the Normans that the Picards mocked

[76] "Historia Cortusiorum," Muratori, *Scriptores*, Vol. 12, cols. 926–927. Neuburger, *Geschichte der Medizin*, II, mentions Genoa and Milan as cities which tried the embargo.

[77] See above, p. 3.

[78] R Maiocchi, *Codice diplomatico dell' università di Pavia*, Pavia, 1905–1915, Vol. 1, No. 263.

[79] Michaelis Platiensis, "Historia sicula ab anno 1337 ad annum 1361," in R. Gregorio, *Bibliotheca scriptorum qui res in Sicilia gestas retulere*, 1791–1792, Vol. 1, p. 562 ff.

at them,[80] but the following stanza, judging from the plural *pestes* in the third line, probably originated in a later outbreak:

To wish the pest to others.
O motive power who movest all that is on earth,
Pray preserve Paduans safe, O Thou Thyself, our Father,
Who livest eternal, on them direct not pestilences,
Let them come upon Venetians, let Saracens too be smitten.[81]

A very full series of ordinances against the plague, issued by the city of Pistoia in May and June, 1348, has been preserved.[82] As there seems to be a general tendency in histories of medicine and works on the Black Death to mention, as the first extant public ordinance of this kind, that issued by Barnabò Visconti in 1374, it seems well to treat the Pistoian enactments in some detail.

There are four ordinances, composed by certain wise men chosen from the people of Pistoia by the *anciani* and the *vexilliferus justitie.* The first was drawn up 2 May, 1348, by Simon Bonaccursi, notary of Pistoia, who was commissioned to see that it be made public in the customary manner. The second and third were drawn up 23 May and 4 June respectively by the same notary, but the fourth, of 13 June, 1348, by another (Francischus Ser Vannis),

[80] H. E. Rébouis, *Étude historique et critique sur la peste,* Paris, 1888, pp. 48–49.

[81] E. Wickersheimer, in *Archiv,* III, 349:

"Pestem optat aliis.
"O motor verus qui moves omnia mundi
Tu sanos Patavos conserves ipse paterque
Qui vivis eternus, nec pestes porrigas ipsis
Hec Venetis veniant, attingant et Sarracenos."

[82] Alberto Chiappelli, "Gli ordinamenti sanitari del Comune di Pistoia contro la peste del 1348," in *Arch. stor. ital.,* Ser. IV, Vol. XX (1887), pp. 3–24. He publishes the complete text of the documents from the Pistoian city archives.

which seems to indicate that in the intervening nine days Simon had fallen victim to the plague.

The decree of 2 May is long, filling nine octavo pages of fine print, and contains full regulations governing protection of the city from contagion from without; precautions against infection inside Pistoia; and measures to avoid frightening the sick. The penalties for violations were fines in all instances. No one was to be allowed to go to the Pisan or Luccan states or return from them, to bring into the Pistoian state any old linen or woolen garments, or to bring corpses into the state. In the markets, dealers in provisions, especially in meats, were to be carefully supervised to insure the quality, preparation, and freshness of their stock; nearly half of the decree is devoted to minute directions regarding them. Attendance at funerals was limited to the family of the deceased, and none were to gather to solace the widow, except when returning from the church or burial, besides her blood kin or four women chosen especially for the purpose. The depth at which bodies were to be buried, and the conditions regarding their removal, were carefully prescribed. To avoid disturbing the sick, bells were not to be rung for funerals, and there was to be no public announcement by crier or trumpeter. Exemption was made in the case of any knight of the court, doctor of laws, judge, or physician, who, as befitted their worth, might be shown at their funerals any honors which were pleasing to their heirs.

Three weeks later a second edict was issued, for the announced purpose of making corrections and additions to the first. It totally annulled the first provision of the latter, that prohibiting the Pistoians from having intercourse with anyone coming from the Pisan and Luccan communities, for the reason, probably, that pestilence was raging so fiercely in Pistoia that it was unnecessary to prohibit intercourse with other places. Changes were made in some of the requirements for butchers and meat sellers; and there was an emphatic reminder of the obligations and powers of authorities for discovery of violations, and for enforcement, of the enactments, strongly reminiscent of governmental admonitions today in connection with our own Gargantuan effort at sumptuary legislation.

About ten days afterward further modifications and addenda appeared. The restrictions in regard to funeral services and the bereaved had evidently not met with success, for, while the ringing of bells was still prohibited, persons might attend funerals by express invitation, and a larger number of relatives and near neighbors were allowed to comfort the afflicted. Sixteen men, at the least, were to be chosen from each district of the city and were to hold themselves in readiness to carry corpses to burial, with no other permitted to do so. They were to be paid and must go when asked; otherwise they would be fined, the oath of anyone who had asked them to come being accepted as full proof against them. The scarcity of wax precluded the use of candles for the dead; instead, from six to twenty-four denarii might be given the priest or friar who officiated. In the case of the poor, payment was left to the discretion of the rector and friars. There was more about discovering violations and enforcing requirements, and another exception was made: if any of the enactments that had been issued were against the liberty of the Church, they were to be null, void, and of no force or value.

The last set of ordinances, that of 13 June, 1348, was to provide for guarding the city and fortified places. In order to spare the cavalry, the more valuable and better citizens of the state, the fatigue of mind and body which had been proved to induce pestilence, the senate made certain provisions. Each cavalryman must either make his prescribed rounds, provide a substitute, or pay a moderate fine. Citizens were to be chosen and paid to keep guard at the gates, and in the strongholds without the city.

As the Theodosian Code gives the best picture we have of the forces working for decay within the Roman Empire, and as the capitularies of Charlemagne depict his age more clearly than do contemporary chronicles, so it would be hard to find a more direct and gripping portrayal of a city stricken by the plague than is unintentionally set forth in the successive Pistoian enactments. The portions bearing most obviously upon our subject are the two exemptions, the one allowing special funeral honors to be paid jurists, judges, and physicians; the other revoking, where the Church was concerned, any regulation to which she objected. Both show the esteem in which intellectual classes were held, but the second is an indication of one of the effects of the Black Death—the heaping of favors upon the Church—which was to react so disastrously later, and which has led some writers to consider the epidemic one of the chief causes of the Protestant Revolt.[83] Another illustration of the same tendency is the official requirement of payment for clergy officiating at a funeral, a payment from which even the poor were relieved only at the discretion of the clergy themselves.

In his decree of 17 January, 1374, Barnabò Visconti issued enactments tending to increase both the mortality of the clergy by pestilence, and the accumulation of property in their hands. The lack of wisdom in his ordinance, and the severity of the penalties, contrast strongly with the careful, public-spirited, and often-revised regulations drawn

[83] See above, p. 4.

up by the wise men chosen from among the citizens of Pistoia. The edict is issued in his capacity of lord of Milan and vicar-general of the Empire, and its object is declared to be the preservation of his subjects "from the contagion of the disease." As soon as the pestilence should appear in any place, everyone was to depart from it immediately into the fields and forests, there to await death or recovery. The priests of the parochial churches were to visit the sick, examine them, and report to officials appointed for the purpose, on pain of the fire. In return, perhaps, for this hazardous service, all the goods of the dead, movable and immovable, must go to the Church. It was also commanded that those who had spread the disease should be deprived of their goods without restitution; that those who had tended the sick must cut themselves off from communication with others for ten days; and that anyone, besides the persons designated above for the purpose, who should care for the sick, was to be punished with death.[84]

Ragusa issued in June, 1377, the first quarantine regulation, as far as is known, which was really a *trentina,* since it provided that all comers from pest-infected regions must stay a month in Mercana or Old Ragusa;[85] six years later Marseilles established as the period of detention the forty-day quarantine.[86]

It seems that, as wars are the diet upon which military establishments thrive, so the Black Death was, on the whole, of benefit to the medical profession. The loss of

[84] "Chronicon regiense," in Muratori, *Scriptores,* Vol. 18, col. 82.

[85] Lechner, *Das grosse Sterben in Deutschland . . . 1348–1351,* Innsbruck, 1884, pp. 65–69, quoting from the *Gymnasial-programm von Zara,* Nos. 24, 32.

[86] Neuburger, *Geschichte der Medizin,* II, 478. He also gives the Ragusa regulation.

leaders in any field of intellectual activity is a blow effects of which are incalculable, and physicians and surgeons of outstanding ability and achievement are numbered in the twenty-five or thirty per cent which our statistics indicate as the approximate mortality among medical practitioners. But it is undeniable that in the last half of the fourteenth century there was a strikingly large number of distinguished physicians and surgeons, and that they were much sought after by rulers and other persons of eminence. Their salaries and fees rose perceptibly, other evidence bearing out Palmieri's observation that the medical art was very lucrative in the second half of the fourteenth century, with physicians of the seventies and eighties, in the region he was studying, leaving large patrimonies, and being the richest proprietors of certain districts which he mentions.[87]

The practice of post-mortems and of anatomical dissection in general appears to have been stimulated by the epidemic, and the upward movement of surgery to have been accelerated. The public was freely instructed, in tractates generously written and scattered abroad, in principles of hygiene, sanitation, and personal prophylaxis, and even so harsh a critic of fourteenth-century medicine as Neuburger recognizes a general improvement along those lines. It was an era of public sanitary ordinances, and marked the beginning of formal quarantine. There was intensive study of the causes of epidemic, leading to many conclusions which are today considered false, such as the relation between heavenly bodies, natural phenomena, and human beings; but also giving a decided impetus to de-

[87] A. Palmieri, "L'arte medica nell' antico Appenino bolognese," in *Atti e memorie della Regia Deputazione di Storia Patria per le provincie di Romagna,* Series 4, Vol. 1 (1911), pp. 257–258. See also Carabellese, *La peste de 1348,* pp. 38–39, and above, pp. 98–99.

velopment of the theories of contagion and infection. While some of the hypotheses advanced, such as infection by the glance of the eye, were wide of the mark, others, for instance the development of immunity, and carriers of disease, have been more happily borne out by the researches of succeeding scientists.

In consideration of both cause and cure, the function and action of heart and lungs, the circulation of the blood, and the increase and spread of toxins in the body were given thought and discussion. The chief interest of treatment in itself centered in maintaining the strength of the heart, in drawing the poison out of the system, and in methods of dealing with imposthumes. The last was largely surgical, and gave excellent opportunity for experimentation, as well as tending possibly to foster specialization. Noteworthy for those who base their ideas about medieval science on long-accepted generalizations, is John of Burgundy's[88] exaltation of modern masters, in the field of epidemiology, over even such giants of antiquity as Hippocrates and Galen. The passage stands near the end of his *Treatise on the Epidemic Sickness,* written about 1365:[89]

[88] It does not seem advisable to treat in this study the physicians, about forty in number, who in the last half of the fourteenth century wrote plague tractates bearing their names; I hope to consider them and their tractates in an article to be published later. John of Burgundy was a professor of medicine and citizen of Liège, who is now generally believed to have been the author of the *Travels of Sir John Mandeville.* The literature on the subject is large; a good starting point is the article "Mandeville" in recent editions of the *Encyclopedia Britannica.*

[89] Sudhoff, *Archiv,* V, 63–69; the passage quoted here occurs pp. 68–69:

"Nec vidi aliquem sub hac forma fleobothomatum, si bene se rexerit et cordis confortativa receperit, quin a mortis periculo evasisset, unde dicere ausus sum, non tamen in obprobrium auc-

And from this I dare say (though not to the disparagement of the authorities, but because long experience makes these things clear) that modern masters everywhere in the world are more skilled in pestilential epidemic diseases than all the doctors of the art of medicine and the authorities from Hippocrates down, however many they are. For no one of them ever saw so general or lasting an epidemic, nor did they test their efforts by long experiment, but what most of them say and treat about epidemics they have drawn from the sayings of Hippocrates. Wherefore the masters of this time have had greater experience in those diseases than all who have preceded us, and it is truly said that from experience comes knowledge.

---

torum, sed quia experiencia longa hoc manifestat, quod magistri moderni ubicunque terrarum in morbis pestilencialibus epidemialibus sunt magis experti quam omnes artis medicinae doctores et auctores ab Ypocrate inferius quotquot sunt. Nam nullus eorum tempore epydimiam vidit regnantem [suo nisi solus Ypocras qui eam vidit regnantem] in civitate Craton [is], quae tamen non multum duravit. Nichilominus tamen super hiis, quae dictus Ypocras vidit, librum edidit de epidemia. Sed Galienus, Diastorius, Rasis, Damascenus, Geber, Mesue, Capho, Constantinus, Serapion, Avicenna, Algagel et sequaces eorum nunquam viderunt ita generalem nec longam epidemiam, nec curas eorum longa experientia probaverunt, sed quidquid plurimi dicant et tractent de epidemiis a dictis Ypocratis exhauserunt. Unde magistri huius temporis in illis morbis maiori experientia usi sunt, quam omnes qui nos praecesserunt, unde dicitur et verum est, quod experientia facit artem."

The phrases in brackets occur in a MS of the fifteenth century, in the possession of Mr. Ludwig Rosenthal, in Münich. The MS here used by Professor Sudhoff is from the Erfurt *Amploniana*, Codex in Quarto 192, fols. 146r–148v, a copy dating, he thinks, from about 1380.

Chapter V

EFFECTS IN OTHER FIELDS OF INTELLECTUAL ENDEAVOR

Astrology, astronomy, mathematics, and physics were in the fourteenth century closely allied with medicine, and were taught in the universities by the medical faculties. Hence the physicians whom we have considered were often proficient in these fields, while investigators especially interested in the principles and phenomena of inorganic matter were likely to be also medical practitioners. Little distinction was made between astronomy and astrology. Those men occupied chiefly with the mathematical side of the subject will be considered here as astronomers, and grouped with mathematicians and physicists; those whose names are connected entirely or very largely with judicial astrology will be in a separate classification.

The importance of the part assigned the heavenly bodies as the superior root of the plague is apparent in the treatises that have been discussed. But the evidence here does not seem to bear out Professor Duhem's findings, based on his wide study of scientific writings of the period, respecting the decided difference between French and Italian points of view. He observes that at Paris a rigorous distinction was made between astronomy and astrology, and that the latter was held in by barriers opposed to it by the Christian faith; the Italian art, on the other hand, knew no such bounds, undergoing rather the influence of

Islamic fatalism, whereby its pretentions were extended beyond every limit.[1] In the tractates of the time of the Black Death, however, the most extreme astrological claims are made by the medical faculty of the University of Paris; the Italian writers lay no such stress upon them, and the two Moslem physicians touch upon them briefly and with decided reservations.

In astronomy the Oxford school, centred at Merton College, enjoyed prestige as the result of its activities at this period.[2] Outstanding names are those of John Aschenden, William Reed, Reginald Lamborn, and Geoffrey of Meaux,[3] each of whom wrote prognostications of evils resulting from one or more of the phenomena of the time: the eclipse of the moon and the conjunction of the planets in 1345, the same occurrences in 1349; and from conjunctions of the planets in 1357, 1365, and 1367.[4] Besides this, tables for the meridian of Oxford, attributed to William Reed, start with radices for 1348, the same year which forms the starting-point of tables calculated for the meridian of Salamanca.[5] Three French astronomers, Firmin of Belleval, Leo the Jew, and John of Murs, each wrote a

[1] P. Duhem, *Le Système du monde*, Paris, 1916, Vol. 4, pp. 184–188.

[2] For data about this group of astronomers, see R. T. Gunther, *Early Science in Oxford*, Vol. 2, Oxford, 1923. Simon of Phares (see below, p. 127, note 15) mentions several French astrologers who went to England to study: such were Bertrand of Bonpas, Peter of Valois, William of Toury.

[3] Geoffrey of Meaux was a Frenchman who seems to have taught at Oxford for about twenty years. Duhem, *Système*, IV, 69–70; A. Wood, *History and Antiquities of the University of Oxford*, Oxford 1792, Vol. 1, pp. 450–451.

[4] Wood, I, 450–451, citing Digby MS 176 of the Bodleian Library. Gunther, II, 58–59, publishes an abstract of the contents of this manuscript.

[5] Gunther, II, 46–47.

Prognostication concerning the Conjunction of Saturn, Jupiter, and Mars.[6]

In mathematics and physics, the only discernible effects of pestilence would be on the lives of workers in these fields. Out of a list of twenty-three astronomers, mathematicians, and physicists who may reasonably be presumed to have been active in their professions in the middle of the fourteenth century, twelve certainly survived the pestilence.[7] Only one is said definitely to have died of it: Thomas Bradwardine, newly chosen archbishop of Canterbury, and considered by Cantor the leading English mathematician of his age.[8] It is doubtless the former distinction and not the latter that is responsible for the preservation of exact knowledge of the date and cause of his death. In the year 1348 Bernard Barlaam, mathematician and theologian, appears to have died, since 4 August, 1348, he was succeeded in the office of bishop of Geraci (Italy) by Simon of Constantinople;[9] and in the same year last mention is made of John of Genoa, papal physician, and perhaps the same John of Genoa who wrote the astronomical works, *Canones eclipsium* (1332) and *Investigatio eclipsis solis anno Christi 1337.*[10]

Besides these three who disappeared during the Black Death, there are half a dozen important mathematicians,

[6] Duhem, *Système*, IV, 38–39.

[7] Albert of Saxony, Dominic of Chivasso, John Chillingworth, John of Saxony, Lewis Charlton, Paul Dagomari, Reginald Lamborn, Richard Swineshead, Simon Bredon, Simon Tunsted, Thémon the son of the Jew, William Heytesbury.

[8] M. Cantor, *Vorlesungen über Geschichte der Mathematik*, Vol. 2, Leipzig, 1913, p. 113.

[9] Tiraboschi, *Stor. d. lett. ital.*, Vol. 5, Pt. 2, pp. 681–689; P. de Nolhac, "Pétrarque et Barlaam," in *Revue des études grecques*, V (1892), 94–99.

[10] See above, p. 94, note 3.

astronomers, and physicists who were productive in the forties but were not heard of afterwards. Two of them, Englishmen at Oxford, were calculators of the Oxford Tables: William Reed, active at that university in 1345, and John Maudith; the last recorded date of the latter is 1340, but there is reason to think that he may have lived at Oxford as late as 1346.[11] Geoffrey of Meaux was possibly author of a discussion of the Alphonsine Tables dated 20 April, 1347; he discussed also the astrological causes of the pestilence in a tractate written during or after its occurrence; there is no later trace of him. The last recorded work of Firmin of Belleval is a treatise on calendar reform composed by order of Clement VI in 1345,[12] in collaboration with John of Murs, like himself an illustrious member of the University of Paris. No date is given in the life of another physicist, Nicholas of Outricourt, after 1348,[13] and nothing is known of the Parisian astronomer, John of Linières, after the middle of the century.[14] It would seem hardly likely that from among these eleven men the only victim of the epidemic was Thomas Bradwardine.

Judicial astrology was a flourishing profession in the fourteenth century, particularly from 1350 to 1400, when successful prognosticators were much in demand by men of affairs, to judge from the brief bibliographical notices in Simon of Phares's *Collection of the Most Famous*

[11] Gunther, Vol. 1, pp. 96, 336; Vol. 2, p. 48. Duhem, *Système,* IV, 72–73.

[12] Duhem, *Système,* IV, 69–70 for Geoffrey of Meaux; IV, 52 for John of Murs.

[13] P. Duhem, *Études sur Léonard de Vinci; ceux qu'il a lus et ceux qui l'ont lu,* Vol. 2, Paris, 1909, p. 11.

[14] Duhem, *Système,* IV, 578–581. Professor Thorndike doubts the validity of Duhem's evidence for John's catalogue of the stars, of 1350.

Astrologers.[15] In the first half of the fourteenth century Simon names fifty-two astrologers, of whom seven were in the service of temporal or ecclesiastical dignitaries; from 1350 to 1400 there are eighty-four, of whom thirty-four held such positions.[16] Conjunctions of planets, comets, eclipses, and earthquakes seem to have occurred with unusual frequency after 1340, and upon these the astrologers made many predictions, two favorites, which rarely failed to come to pass, being of another outbreak of pestilence, or of another disaster to the French at the hands of the English. Not only was the plague foretold, but its connection with the heavenly bodies was at times carefully set

[15] E. Wickersheimer, editor, *"Recueil des plus célèbres astrologues et quelques hommes doctes" faict par Symon de Phares du temps de Charles VIII*^e, Paris, 1929.

Simon includes among those who have made successful predictions several cardinals, and some avowed opponents of astrology, like Petrarch and Henry of Hesse. He makes occasional errors of dating, as in putting Roger Bacon in the middle of the fourteenth century; and in a few instances the same man is given in two places, with slightly differing names—*Petrus de Ebano* and *Petrus Apponus,* or *John Essenden* and *John Vischindem.* But after making allowances for errors which might be expected in a work written at the end of the fifteenth century when dealing with men living more than a hundred years before, we get from Simon valuable information about fourteenth-century astrologers.

[16] Simon of Phares was himself a court astrologer, and so was likely to know his predecessors in that office for the past century or more. He gives the following favorites of fourteenth-century princes:

Philip VI of France: Symon of Canteleu.
Charles V of France: André of Sully; Dominic of Chivasso; Gervais Chrestien; Thomas of Pisa.
Charles VI of France: Alexis Volant; Denis of Vincennes; James of Montciclat (attached to Isabel of Bavaria, wife of Charles VI).
Richard II of England: George Scorf.
Leopold of Austria: Firmin of Belleval.
Charles of Blois: Michel of St. Mesmyn.
Bertrand du Guesclin: Yves of St. Branchier.
House of Orleans: Gencien of Beaugenci; Gilbert of Chateaudun.

forth; a detailed illustration of this occurs in a tractate written toward the close of the century, in which every phase of the disease is related to aspects of the planets or some zodiacal calculation.[17]

The prominence of astrologers in the service of the great led to their becoming confidential advisers of their patrons, and being sent on missions of political importance.[18] We read also in Simon of two astrologers who gained favor respectively with Charles V and Charles VI by foretelling uprisings of the common people, of another who prognosticated to Henry IV of England a conspiracy against him, and of one, an Italian, so marvelously expert that no criminal could escape detection by him.[19] It is not possible to draw any conclusions as to the number of astrologers who died of the plague, since data about them are indefinite and sketchy, but it seems that to their profession, as to that of medicine, it brought increased emolument and esteem.

[17] Printed by Sudhoff, *Archiv*, XVI, 96–102.

[18] Some of those given by Simon are:

Anthony of Montpharon, sent to England and Rome by Pope Innocent VI.

Bertrand of Bonpas, sent on mission to Paris by Innocent VI.

William of Toury, who, when sent to England to study because of his knowledge of the stars, went the more willingly in order to relieve the tedium of the captive King John.

Denis Plusdoré, sent to England to bring back to France Isabel, widow of King Richard.

Gencien of Beaugenci, sent on mission by the duke of Orleans to Pope Benedict (chosen anti-pope 1394), whom he inclined to the Orleanist side.

Espinel de la Mirandolle, chosen by Reggio to do honor to the countess of Touraine.

Simon of Perre Couvert, chosen by the people of Ghent to remonstrate on their behalf with Count Louis.

[19] Anticipating the psychometrists, or clairvoyant detectives, who, according to Sir Conan Doyle, will be part of every well-equipped police station of the future. *New York Times*, 17 August, 1929, p. 2.

A striking feature about the last half of the fourteenth century is the greater amount of lawlessness then prevalent, and the number of outbreaks, both popular and intellectual, against authority. The popular revolts lie outside our sphere, except in so far as they may have been incited by men of learning, but the question arises of the effect the pestilence may have had upon those active in public affairs, political or legal.

Deaths credited to this plague from among political thinkers are those of Alfonso XI, enlightened ruler of Castile, who succumbed while besieging Gibraltar, and of one of his foes, the Moor Muhammad ibn Ahmed Alansari, who had fled before the Christians, and was in Ceuta writing a book advocating a war against them when the plague overtook him.[20] In England John Stratford, archbishop of Canterbury, a politician whom Stubbs calls "the most powerful adviser of the constitutional party," died in 1348,[21] and it was between 1347 and 1349 that we lose sight of William of Occam, writer on politics, philosophy, logic, and theology.[22]

Certain historians or chroniclers interested in public affairs either died, or at least abruptly ceased writing at the time of the pestilence. Giovanni Villani,[23] the Florentine historian, and Buonincontro Morigia, of Monza in Lombardy, each not only wrote a history of his fatherland, but took an active part in public affairs. Of the former it is definitely stated at the conclusion of his *Istoria* that he died of the Black Death in 1348; [24] of the latter there is no

[20] Casiri, *Bibliot. arab.-hisp. escurial.*, II, 89.

[21] *Dictionary of National Biography*, LV, 30–33.

[22] DNB, XLI, 357–362.

[23] Philip Villani, *Le vite d'uomini illustri fiorentini*, Florence, 1826, p 58.

[24] Muratori, *Scriptores* Vol. 13, cols. 1001–1002.

record after 1349.[25] Boniface Morano wrote a chronicle of Modena from 1306 to 1342; an inscription on his tombstone, preserved there in the church of St. Francis, showing that he died in 1349, is quoted by Muratori, though he is not sure of its genuineness.[26] and an anonymous contemporary Pistoian chronicle, beginning in 1300, breaks off abruptly in 1348.[27] Two German historical works make the middle of the fourteenth century a dividing point: the Magdeburg chronicle of Henry of Lammespringe, who chooses the year of the great mortality, 1350, as the beginning of his last book, the contemporary portion of his work;[28] and an Augsburg chronicle, from 1348 to 1477, which includes the burning of the Jews and "a great epidemic" among the brief items of its first year.[29] Höniger, though in general inclined to minimize the effects of the plague, calls it a striking fact that in two-thirds of the German chronicles dealing with the period there is a complete gap for the sixth and early seventh decades of the fourteenth century; and that the others, except the Austrian sources, offer only brief and unimportant comments.[30]

A famous legist who fell victim to the plague was John Andrea, whom Tiraboschi calls the most celebrated canonist not only of his own century, but perhaps of any time. He was professor of canon law in Bologna and in Padua,

[25] Muratori, Vol. 12, col. 1061 ff.

[26] Muratori, Vol. 11, p. 92.

[27] Muratori, Vol. 11, col. 359 ff. It is to this chronicle that the Italian version of the Parisian *Consultation* is appended. See above, p. 16, note 33.

[28] "Die Magdeburger Schöppenchronik Heinrichs von Lammespringer," edited by K. von Janicke, in *Die Chroniken der niedersächsischen Städte*, Vol. 7, Leipzig, 1869, pp. 218–9.

[29] Same series, Vol. 22, Leipzig, 1892, edited by Fr. Roth.

[30] Robert Höniger, *Der schwarze Tod in Deutschland*, Berlin, 1882, p. 3.

wrote many works on the subject, and was the father of two daughters, Novella and Bettina, who were prominent in learned circles of Italy.[31] A younger canonist than John, Achino Orsi Carnelli of Chioggia, was elected professor of canon law at Padua immediately after receiving his doctorate there with distinction in 1348, but in the same year died of the plague.[32] Epitaphs from the tombs of these two men lament the loss the profession sustained in their deaths. Other legists whose death is directly attributed to the Black Death were the first lay chancellor of England, Robert Bourchier, or Boussier, former chief justice of Ireland and five times a member of parliament;[33] Mattagnano, or Mascagnano, degli Azzoguidi, writer on civil law, son-in-law of Taddeo Pepoli, and only twenty-five at the time of his death;[34] and, in Granada, an author of a long work on canon law, Alkaluzi, and a teacher of jurisprudence, Aliatin. There died during the epidemic two other Moorish jurisconsults, Ibn Algiab and Albadui,[35] and also an aged Italian doctor of civil and canon law, Franciscus Barbaremus.[36] Guy Guisi, doctor of law and bishop of Concordia, died in 1347,[37] and another bishop and lawyer, Guy son of Philip of Baiso, in 1349.[38]

The distinguished jurist, Bartolus of Sassoferrato, was in 1343 called from the chair of law which he occupied at the University of Pisa to a similar position at Perugia, and given citizenship of that city in 1348. A law forbid-

[31] Tiraboschi, V, pt. 2, 522–540; Gloria, "Univ. Pad.," in *Istituto Veneto, Memorie,* XXII, p. 566.
[32] Gloria, *Università di Padova,* I, 310–311.
[33] DNB, VI, pp. 14–15.
[34] Tiraboschi, V, pt. 1, 512–513.
[35] Casiri, *Bibliot. arab.-hisp. escurial.*, II, 83, 85, 110, 84.
[36] Villani, Philip, *Vite,* pp. 36–39.
[37] Tiraboschi, V, pt. 2, 521.
[38] Tiraboschi, V, pt. 2, 520.

ding Perugian citizens to teach in the university there was waived in his favor, and he filled the position with distinction to the institution and himself till his death nine years later.[39] In his *Commentary* on the Digest, he speaks of the effect of the Black Death on law: Prescription and usucapion [40] do not hold good in time of war, if it is so great a one that laws are not administered in the state, or in time of mortality. Illustrations are given of wars during which this occurred.

> Now, for instance, from the same equity, it might be observed that in the time of the urgent mortality of the year of our Lord 1348, as you know, the pestilence was so great that laws were not administered in the states, and an infinite number of men died, that in that time usucapion slumbered . . . and that the hostility of God was stronger than the hostility of men.[41]

A hundred years later, in England, Thomas Gascoigne observed that "before the great pestilence in England there were few quarrels among the people, and few lawsuits; and so there were also few lawyers in the kingdom, and few in Oxford."[42] After another century of intermittent out-

[39] For a recent study of Bartolus, see C. N. S. Woolf, *Bartolus, His Position in the History of Medieval Thought*. Cambridge, 1913.

[40] The acquisition of title to property by the uninterrupted possession of it for a certain time, if acquired in good faith. Usucapion dates back to the old Roman *Jus Civile*, prescription is praetorian.

[41] Bartolus de Saxoferrato, *Commentaria*, Venice, 1590–1602, Vol. 5, pp. 91–92: "Idem forte dicendum esset ex eadem equitate, quod tempore mortalitatis instantis de anno Domini 1348, prout scitis, erat tanta pestilentia, quod jura non reddebantur in civitatibus, et moriebantur infiniti homines, quod tempore illo usucapio dormiebat, arg. d. c. ex transmissa et fuit hostilitas Dei fortior quam hostilitas hominum." I am indebted for this reference to Miss Beatrice Reynolds, a member of the history department of the Connecticut College for Women.

[42] Thomas Gascoigne, *Liber veritatum*, written 1433–1457. J. E. T. Rogers (ed.), *Loci e libro veritatum*, Oxford, 1881, p. 202:

bursts of plague, the Florentine jurist, Sylvester Aldobrandini, wrote a *Treatise on the Pestilence and Its Effects on Law.*[43] Thomas Gascoigne ascribed the quarrels and increased litigation in his time to the difference in the quality of the clergy before and after the great pestilence. Writers of the second half of the fourteenth century had expressed sentiments of the same kind, and much has been written in comparatively recent times about the effect of the Black Death on church and clergy. Seebohm and Gasquet, after careful studies of the mortality among the English clergy, based on episcopal and parish registers, each concluded that more than half the beneficed clergy died in 1348 and 1349.[44] In German works dealing with the Black Death, estimates are given of effects upon the

"Si enim episcopus debite curae suae diocesis vacaret in singulis diocesibus, non esset multiplicacio tot malorum, pro quibus reformandis jam necesse est recurrere ad papam, ad regem, ad consilia, et ad parliamenta. 'Obsta principiis, cito medicina paratur'; ante enim magnam pestem in Anglia paucae fuerunt querelae in populo, et paucae implacitaciones, et sic fuerunt item pauci legistae in Angliae regno, et pauci legistae in Oxonia, quando fuerunt triginta milia scolarium in Oxonia, ut vidi in rotulis antiquorum cancellariorum Oxoniae, quando ego fui ibidem cancellarius; et promocio bonorum virorum, et residencia in suis curis, et ecclesiae sufficienter dotatae et non appropriatae aliis non curatis, extra parochiam manentibus, fuerunt causae quare paucae lites tunc regnabant in parochiis, et pauci errores in comparacione litium et querelarum et errorum qui jam sunt."

[43] Vatican *Codex latinus* 5843: "Incipit utilis Tractatus de Peste, et ejus effectibus in jure editus per eundem Dominum Sylvestrum Aldobrandini in felici Pisano Gymnasio." This MS, which has never been published, was called to my attention by Professor Lynn Thorndike.

[44] F. Seebohm, "The Black Death and Its place in English History," *Fortnightly Review,* Sept., 1865, pp. 149–160; F. A. Gasquet, *The Great Pestilence,* London, 1893, later published as *The Black Death,* 1908.

German clergy;[45] statements of opinion, apparently not based on first-hand investigation, have been made from time to time by others interested in the subject.[46] There are to be found attributed to it, besides great mortality, the consequent enforced ordination of men and youths of inferior qualifications; general demoralization of both regulars and seculars; the wiping out of existence of practically the whole personnel of some monasteries, and the closing of many parish churches. Further consequences are said to have been great multiplication of plural benefices, and increase in the property of the church to the extent of arousing the apprehension of secular powers. The Black Death has been called an important cause, even the main cause, of the Reformation. The data given here are offered in the hope that they may shed some further light upon the matter.

An examination of Eubel's *Catholic Hierarchy*[47] makes it possible to estimate with more accuracy than in other professions the number and percentage of deaths among the higher clergy. Death is assumed only in cases where it is positively stated, with either the date of death or, when this is not given, the date of the confirmation of the successor in office coinciding with the dates of the Black Death in the particular locality where the incumbent was at the time. This was usually 1348 and 1349, though the latter part of 1347 is included for certain Mediterranean lands, and 1350 for some of the north German countries

[45] Heinrich Haeser, *Lehrbuch der Geschichte der Medizin und der epidemischen Krankheiten,* Vol. 3, 3d ed., Jena, 1882, pp. 97–156; Robert Höniger, *Der schwarze Tod in Deutschland,* Berlin, 1882; Karl Lechner, *Das grosse Sterben in Deutschland,* Innsbruck, 1884.

[46] For general estimates of the importance of the Black Death, see above, pp. 4–5.

[47] Conrad Eubel, *Hierarchia catholica Medii Aevi,* Vol. 1, Münster, 1913.

and for Russia.[48] Of course the appointment or confirmation may not have been given till a year or more after the death of the former holder, but the same thing might be true of similar documents issued in the early fifties, so these divergencies probably offset each other. In cases where the incumbent disappears without statement of death, and his successor is appointed or confirmed during the Black Death period, the presumption is in favor of death, but the cases are listed separately. Those places are omitted where gaps occur or dates are not given for the late forties; metropolitans and bishops listed as "survived" are incumbents of metropolitanates or dioceses in which no death is noted while the plague was prevalent there.

There were 27 or 28 cardinals who held office during the period of the Black Death; 7 died between May and August, 1348, one in October, 1349, and in the case of one it is doubtful whether his death occurred in 1344 or 1348.[49] Since there is some overlapping of cardinals with other prelates the former are omitted from the table which follows. Eubel gives 5 patriarchs at this time, of whom one, Peter, patriarch of Jerusalem, died and was succeeded in office in March 1349. The proportion of deaths among metropolitans and bishops was even higher than that among cardinals: of 70 metropolitans, 25 died in 1348 or 1349; one, metropolitan of Salerno, in 1347; and one presumably in 1349, since he had disappeared and his successor had received papal confirmation in June 1349. Two hundred seven bishops died while the Black Death was prevalent in the lands in which their sees were located, and during the same period were confirmed the successors of 13 bishops

[48] See above, p. 1.
[49] Eubel, Vol. 1, pp. 15–18.

who had disappeared. In 368 dioceses the incumbents survived, no deaths being recorded between the fall of 1347 and the end of 1349. The results may be tabulated:

	Died during Black Death	*Disappeared, Successor Appointed during Black Death*	*Doubtful for Other Reasons, as Borderline Dates*	*Survived, Holding Office through Black Death*
Popes........	0	0	0	1
Patriarchs....	1	0	0	4
Metropolitans	25	1	0	44
Bishops......	207	13	20	368
Total......	233	14	20	417

Since those who disappeared and whose successors were appointed during the Black Death may reasonably be assumed to have died, there are 255 deaths to 435 survivals, or a mortality of 36 per cent. Or, omitting the 14, the mortality is still over 35 per cent.

Comment upon the havoc wrought by the epidemic in the ranks of the clergy is made by the contemporary author of the *Second Life of Clement VI:*[50]

> For on account of the aforesaid losses of the mortality, he [Clement] renewed, as it were, all the churches of the world in prelates and rectors. And he restored in a praiseworthy manner even the college of the lord cardinals, it having been left greatly diminished from the aforesaid mortality which absorbed many of them, reverend men indeed, conscientious and learned.

[50] E. Baluze, *Vitae paparum avenionensium, 1305–1394,* Paris, 1914–1922, Vol. 1, p. 270. This is repeated in the "Third Life of Clement VI."

In England as early as 1348 an anonymous contemporary chronicler remarks that there is such a scarcity of priests and others serving churches that many of the latter remained without incumbents.[51] The first indication of the plague in the Patent Rolls of Edward III is a royal signification in February, 1349, to the bishop of Salisbury that a canon of the monastery of Ivychurch, where there had been a college of thirteen canons, has brought the king news of the death from plague of the prior and all the canons but himself; and that Edward has "given the royal assent to him as if he had been elected to be prior."[52] From then till February, 1350, occur a succession of licenses to elect and presentations to benefices because of voidances of sees and abbeys, and also because of the war with France.[53]

In the report of the Historical Manuscripts Commission, *The Dean and Chapter of Canterbury*, there are comments upon the large number of presentations to benefices bearing the date 1349;[54] there were ten times as many vacancies during the pestilence as occurred over the same period of time twenty years later.[55] There is a copy of a letter to the pope, dated 8 February, 1350, imploring him to free the see of Canterbury from the dues customary upon the vacancy and refilling of the see, because of the great loss suffered from the pestilence of the two preceding years.[56] At almost exactly the same time, 12 February, 1350, Ed-

[51] J. Tait (ed.), *Chronicon anonymi cantuariensis*, Manchester, 1914, p. 193.

[52] *Calendar of Patent Rolls*, VIII, 260.

[53] *Patent Rolls*, VIII, 251–479.

[54] *Historical Manuscripts Commission*, 5th report, appendix, 1874, "Dean and Chapter of Canterbury," C. T. Martin (ed.), p. 430.

[55] *Hist. Mss. Comm.*, 8th report, appendix, 1881, "2nd report on MSS belonging to the Dean and Chapter of Canterbury," p. 338.

[56] Same, Appendix to 8th report, p. 341.

ward III granted that "in consideration of the intolerable charges sustained by the abbot and monks of St. Albans by frequent voidances of their abbey" they might pay him in instalments the thousand marks due for the last voidance.[57]

John of Reading, a contemporary chronicler, complains bitterly, under the year 1349, of the Mammon of unrighteousness which has taken possession of the regulars, especially the mendicants, who, having become unduly rich through confessions and legacies, were seeking after earthly and carnal things.[58] But the most scathing charges concerning the demoralization of the clergy came from two archbishops of Canterbury, mandates issued in 1350 and 1362, respectively, by Simon Islip, and in 1378 by Simon of Sudbury, to compel chaplains to serve church cures and to accept reasonable salaries. The first pronouncement, that of 1350, declares that priests who have survived the past pestilence display insatiable avarice, charging excessive fees, and neglecting the care of souls.[59] The second and third make the same charge, and add that the priests now desire voluptuous pleasures to such an extent that souls are neglected and churches and chapels are empty, to the horror and scandal of churchmen, and acting as an evil example to laymen.[60]

[57] *Patent Rolls,* VIII, 476.

[58] J. Tait (ed.), *Chronicon Johannis de Reading,* Manchester, 1914, pp. 109–110.

[59] D. Wilkins, *Concilia Magnae Britanniae et Hiberniae,* London, 1737, Vol. 3, pp. 1–2.

[60] Wilkins, III, pp. 50–51, 135–136. The wording of a portion of the charges seems better left in the original Latin: "sacerdotesque praedicti sic delicati et cupidi salariis excessivis hujusmodi farcinati evomunt, indomiti delirant et deficiunt, et post ventris ingluviem, et varias carnis illecebras, spumant in libidinem; et tandem, in malorum voraginem penitus demerguntur in virorum ecclesiasticorum

It is difficult to determine much about the effects of the Black Death in France, especially where social and economic factors were involved, because they became blended with the disastrous consequences of the Hundred Years' War.[61] Instances are encountered where the special cause is alleged, such as the countess de l'Isle's representations to Philip VI, in April, 1350, that the abbey of Villelongue in Carcassone was so impoverished by the mortality that the monks could not live unless they were allowed to keep the property bequests frequently made them by the devout for more than the year and a day to which they were limited.[62] A more satisfactory view of the consequences of the mortality upon church and clergy can be obtained, however, in countries where the issue was not thus complicated.

The papal documents and registers from 1295 to 1352 having to do with the territory comprised in the present district of Saxony and its environs give an opportunity for comparison of mortality among the clergy for the four-year period before the Black Death, 1344–1347, with the period 1348–1351.[63] Since the documents cited usually announce the appointment of a new holder of a benefice, the death of the former incumbent may or may not have taken place in the same year. From 1344 to 1347 only

---

detestabile scandalum, et exemplum perniciosum laicorum. Huic igitur morbo pestifero mederi salubriter et celeriter cupientes. . . ."

[61] H. Denifle, *La Guerre de cent ans et la desolation des églises, monastères, et hôpitaux en France*, Paris, 1897–1899, practically ignores the part played by the pestilence.

[62] Devic et Vaissette, *Histoire générale de Languedoc*, Vol. 10, col. 1051–1052, citing Archives nationales JJ. 78, n. 244.

[63] Gustav Schmidt, *Päpstliche Urkunden und Regesten aus den Jahren 1295–1352, die Gebiete der heutigen Provinz Sachsen und deren Umlande betreffend.* Issued by the Historical Commission of the District of Saxony, Halle, 1886.

three deaths are noted; from 1348 to 1351, twelve. The number of vacancies in the archbishopric of Canterbury at this time[64] is rivaled by those in the bishopric of Naumberg. In January, 1349, the pope appointed his chaplain, Nicholas, to that bishopric in the place of Widego of Ostrau, who had died 24 October, 1348. Two years later, 22 October, 1350, Nicholas received a further promotion, this time to the patriarchate of Aquileia, left vacant by the death of Bertrand of St. Gênes, and the following July Clement confirmed the election of Canon John of Miltitz to the bishopric of Naumberg. John died in less than six months, and the bishopric was then given to a Breslau canon, John of Neumarkt; the latter, however, did not enter upon it, but instead Rudolf Schenk of Saalect, in the same year, 1352.[65] Clement's document conferring the patriarchate of Aquileia begins:[66]

Avignon, 25 October, 1350

Observing with the keen scrutiny of solicitous consideration how burdened with losses, how full of perils is the daily voidance of churches, and what great dangers it involves, we gladly apply our hands, already full of labors, and bestow fervent endeavor to place over the widowed churches suitable pastors, and to send prudent assistants, as the quality of the time requires and the suitability of the places indicates, sometimes through the employment of due provision, sometimes indeed by means of appropriate transferral, . . .

[64] Three archbishops of Canterbury died within a year: John Stratford, 23 August, 1348; John Ufford, 20 May, 1349; Thomas Bradwardine, 26 August, 1349. The dates are Old Style, as they appear in a memorandum written shortly after Thomas Bradwardine's death, and preserved among the manuscripts of the Dean and Chapter of Canterbury. *Hist. MSS Comm.*, Appendix to the 8th report, 1881, p. 337.

[65] G. Schmidt, *Papst. Urkunden u. Regesten*, documents 161, 190, 205, 212.

[66] Schmidt, document 190.

From the same collection of documents an instructive comparison may be made of the holding of plural benefices in Saxony in the three years preceding the plague with that of the three years following it. In the years 1345–1347, inclusive, 39 benefices were held by 13 men; from 1350 to 1352, 57 were in the hands of 12 holders, an increase of 58 per cent.

Registers and archives of monastic establishments give useful information also.[67] The registers of the Benedictine abbey of Pfäffers and of the district of Sargans contain a proclamation by the Abbot Hermann, issued in March, 1350, that an anniversary service is to be celebrated yearly for "all the men, women, and boys of our monastery who in the year of the Lord 1349 died in that great and unheard-of epidemic, whose number extended itself to 2,000 and more." The same records show during the next quarter of a century the incorporation by episcopal decree of various churches into the monastery: the parochial church of Maemidorf, the parochial church of Mails and its two daughter chapels, and the church in Ragaetz.[68] The reasons given are the good works and hospitality of the monastery, and its burden of debt. To the monastery at Interlaken, because of its small income, was united in 1351 the church of Zweisimmen.[69]

The archives of the monastery of S. Giacomo and those of the convent of Sant' Agnese, both at Bologna, show a decided increase in the number of disputes and lawsuits

[67] See P. G. Mode, *Influence of the Black Death on the English Monasteries*, Chicago, 1916.

[68] *Die Regesten der Schweizerischen Eidgenossenschaft* (Mohr, ed.), "Die Regesten der Benedictiner-Abtei Pfäffers und der Landschaft Sargans," Nos. 196, 202, 262, 263.

[69] *Schweiz. Genoss.* (Mohr, ed.), "Die Regesten der Männerklosten zu Interlaken," No. 351.

in which they figured in the twenty-five years following the plague as compared with the same period preceding it. From 1322 to 1346 the archives of both establishments show five documents devoted to the ordinary business of purchases of land, repairs, and rents, seven wills containing bequests, and three concerning disputes or compromises. From 1347 to 1371 there are six of ordinary business, nine wills, and thirteen dealing with disputes or compromises. The records of another monastery of Bologna, that of S. Stephan, are very full for the twelfth and thirteenth centuries, but so scanty for the fourteenth that no conclusions can be drawn.[70]

In a *History of the Swabians* written more than a century after the Black Death the author, Felix Fabri, speaks at some length of the evil consequences of the pestilence of 1348 upon the monasteries in that locality. The mortality was very great, and many monasteries were without monks. "For those who survived were not in the monasteries but in the cities, and, becoming accustomed to worldly ways of living, fell from bad to worse; and there occurred a grave decline of all religions in this pestilence." At that time the monks of the convent of Auwa went from Auwa to Ulm, and tarried there. Since they were very rich and had little religion, they attracted many companions and spent the goods of their monastery in convivial living. Many of their fellows from other monasteries joined them in their feasting, while the temporal goods of their establishments fell into decay. To so low an

[70] The archives of these three monasteries appear in the first three volumes of the *Chartularium studii Bononiensis*, Bologna, 1913. Those of S. Giacomo are in Vol. 1, pp. 191–391, and are edited by Luigi Nardi; those of Sant' Agnese in Vol. 2, pp. 1–197, edited by Albano Sorbelli; those of S. Stephan occupy the whole of Vol. 3.

estate did one monastery fall that at that time it had neither abbot, provost, nor monks, and the convent of nuns there failed completely, "so that today the remains are scarcely visible." A wealthy widow, moved by the desolation of the monastery, stretched out helping hands, and its occupants were gathered together again. "But the observance of the ceremony of the order and rule was not at all maintained. Moreover, as the convent was richly endowed, there arose many dissensions in the elections of the abbots, since many aspired to the dignity; and contentions and suits developed between abbots and monks, whence certain enormities resulted, on account of which the convent fell into great infamy." However, it had recovered, and at the time at which Felix was writing it was distinguished above neighboring monastic establishments.[71]

While this was written too long after the epidemic to be taken at its face value, it is evident that the scandals of that unhappy period had passed into local tradition, and that there still existed more material evidence in the shape of ruins of some of the religious foundations. A contemporary protest against the vices of the clergy was uttered by Ludolph of Saxony who, in 1348, took religious vows and devoted himself to writing a *Great Life of Jesus Christ.* In the early part of the book he calls attention to the fact that John's words to the soldiers, "Content yourselves with your pay," apply to all rectors, all clerics, who exact more than they should.[72] Toward the end he has become more drastic:

[71] Felix Fabri, "Historia Suevorum," *c.* 1480, Book 2, pp. 309–310, in M. Goldast, *Suevicarum rerum scriptores aliquot veteres,* Frankfort, 1606.

[72] Translated into French and edited by M. P. Augustin, Ludolphus de Saxonia, *La Grande vie de Jesus-Christ,* 4 Vols., Paris, 1864–1865. This citation is from Vol. 1, ch. 18, p. 380.

How many there are, alas! who have the title of shepherd and do not blush to play the shameful role of wolves. Far from laboring to make perfect those put under them, they oppose their progress by neglect and bad examples. They send men commissioned to take tithes and confiscate property, or indeed they depute inquisitors, and cause deplorable schisms. And that is why heresy and vice grow and increase in the garden of the church, to such a point that it becomes almost impossible to uproot and destroy them.[73]

In this work Ludolph makes no direct reference to the plague, but shows an inclination to use the word "pestilence" to symbolize the evils of life. He explains that when God says "I who make peace and create evil," He does not mean evil in the sense of sin, "but rather, indeed, pestilence, famine, and other punishments which most people regard as evils,"[74] Again, he reflects that:

The majority of men promise themselves a long life because they are young, of good constitution, strong, and robust. The fools! They know not what may happen tomorrow, and do not consider that unexpected accidents, fever, pestilence, and other maladies to which creatures are subject, carry off many more of them than natural death, and they imagine that they will die while enjoying good habits of body.[75]

He is also prone to medical similes, some of them strongly reminiscent of plague literature. Later, he says, Lucifer will attack Jesus openly, "by the organ of the Jews, in himself and in his members";[76] or, "Neglecting thus to disinfect the source which is poisoned, we make an effort to purify the streams of our maladies which flow from it."[77]

[73] *La Grande vie*, Vol. 3, ch. 51, p. 53.
[74] *La Grande vie*, Vol. 2, ch. 38, p. 361.
[75] *La Grande vie*, Vol. 1, ch. 20, p. 431.
[76] *La Grande vie*, Vol. 2, ch. 22, p. 60.
[77] *La Grande vie*, Vol. 2, ch. 30, p. 180.

The word "corruption" occurs in passages suggestive of contagion: "they are to live in the midst of the world of sinners without contracting any corruption in their society";[78] and elsewhere, "in the desert of a holy life, we live in contact with corrupt men, without our souls' sharing in this corruption."[79]

This, like the pronouncements of the archbishops of Canterbury, Clement VI's introductory remarks when conferring the patriarchate of Aquileia, and Henry Bouhic's thankfulness at escaping the Black Death and so being able to complete his commentaries, is an illustration of one of the most important and possibly far-reaching consequences of the Black Death: its effect upon men's minds.[80] The tendency of the various citations and estimates given here is to bear out the conclusions others have reached of the disastrous results of the pestilence upon the church. It is of especial interest to note, in the calculation based upon Eubel, that the death rate among the higher clergy appears to have been about 35 per cent, since among them were included a large number of the Christians eminent in all the fields of learning. This is about the percentage of mortality among the learned professions indicated throughout this study, where it was possible to obtain fairly accurate and complete data about any particular group, such as the physicians and surgeons of Clement VI, or Ibn al-Khatīb's list of the Arabic physicians of Granada. Since the higher clergy were presumably not so much exposed to the pestilence as were the lower clergy and the rank and file of the population, the figures would seem also to support the estimate of one-half more nearly than that of one-fourth as the population loss sustained by Europe in consequence of the Black Death.

[78] *La Grande vie*, Vol. 3, ch. 58, p. 191. [79] *La Grande vie*, Vol. 2, ch. 22, p. 41. [80] See below, pp. 171–174.

Chapter VI

EFFECTS UPON UNIVERSITIES AND EDUCATION

There has probably never been a time when institutions of higher learning received such undivided allegiance from the intellectual world, and such whole-hearted support from authorities in church and state as did the universities of Europe in the century preceding the Black Death. Because of the attention of prominent men and women thus focused upon them, and in the nature of the case, since they were peculiarly the homes of learning, there are enough documents and references relating to them to give an idea of the effect which the epidemic of 1348 had upon them, and, to some extent, upon education in general. It brought about, in the main, a marked decline immediately after its occurrence or within the two succeeding decades, shown in statements made by the universities themselves, by rulers spiritual and temporal, and by eminent writers, and appearing also in various sorts of indirect evidence. A beneficial result was the effort to counteract the consequences of the pestilence, and of other destructive forces, by the foundation of additional colleges and universities, and by the extension of university privileges. Shortly before the outbreak of the plague, there existed in Europe about thirty universities;[1] of five of them traces disap-

[1] Bologna, Paris, Oxford, Montpellier, Reggio, Modena, Cambridge, Vicenza, Palencia, Padua, Naples, Vercelli, Orleans, Angers, Toulouse, Salamanca, Court of Rome, Piacenza, Arezzo, Seville, Lerida, Rome, Avignon, Perugia, Treviso, Cahors, Verona, Grenoble,

peared in the forties and fifties,[2] but before the end of the century fifteen others arose, six more than had been established in the preceding fifty years.[3] Further effects, or possible effects, were the growth of student rights, university participation in political affairs, changes in secondary education, and increasing use of vernacular languages in the field of learning.

The first alarm of the plague evoked action from the medical faculties of some of the universities, or from their members. In the spring of 1348 Gentile of Foligno, who was then connected with the University of Perugia, and "the venerable college of masters of Perugia" coöperated in issuing directions to the public for prevention and treatment of the epidemic.[4] The following October was promulgated the more ambitious effort of the medical faculty of Paris, a treatise whose origin and purpose are clearly set forth in its introduction:

> All the masters, individually and collectively, of the College of the Faculty of Physicians at Paris, aroused at the command of the most illustrious prince and lord, Philip, king of the French, desiring also to strive for the public good, purpose to set forth in a brief compendium the causes of the epidemic, universal and remote, particular and near, and also the more excellent remedies.[5]

Seven months later great Montpellier, chief center of the medical lore of fourteenth-century Europe, added its contribution in the shape of the tractate written by "a

---

Pisa, Valladolid, Lisbon-Coimbra. See the list of universities in H. Denifle, *Die Entstehung der Universitäten des Mittelalters bis 1400*, Berlin, 1885, pp. 807–810.

[2] See below, p. 162.

[3] Denifle, *Entstehung der Univ.*, pp. 807–810.

[4] See above, p. 12.

[5] H. Denifle, and A. Chatelain, *Chartularium Universitatis parisiensis*, Paris, 1889–1897, Vol. 2, No. 1159. See above, pp. 14–17.

certain physician of Montpellier" and inscribed "to the flourishing medical *studium* at Paris and to the whole university."[6] This inscription is the only passage in the document that may be taken to indicate connection between its author and the University of Montpellier, and the fact is not proved that the "venerable college of masters of Perugia" was part of the university there; but it seems probable that these two works, like the Parisian *Consultation,* were connected with the universities of the towns which gave them birth.

In the establishment of universities, and in the privileges and decrees concerning them, we find a few direct statements of the effects of the pestilence, beginning before it ceased and extending through the succeeding twenty years or more. In 1348 the canny Florentines determined to make their *studium* a *studium generale,* setting their plans afoot, according to Villani, as soon as the plague relaxed, with the intention of attracting people to the city, of increasing it in fame and honor, and of fostering learning and virtue among the citizens.[7] Carabellese considers this a pretext, the real object being to remedy the depopu-

[6] See above, p. 21.

[7] Matteo Villani, "Istoria," Book 1, ch. 8, in Muratori, *Scriptores,* XIV, col. 18: "Rallentata la mortalità, e rassicurati alquanto i Cittadini, che haveano a governare il Comune di Firenze, volenda attrare gente alla nostra Città, e dilatarla in fama, e in honore, e dare materia a suoi Cittadini d'essere scienziati, e virtuosi, con buono consiglio, il Comune provvide, e mise in opera che in Firenze fosse generale Studio di catuna scienzia di legge Canonica, e Civile, e di Teologia. E a ciò fare ordinarono uficiali, e la moneta, che bisognava per havere i Dottori delle scienzie, stanziò che si pagasse annualmente della Camera del Comune, e feciono acconciare i luoghi dello Studio in sù la via, che traversa da casa Donati, a casa i Visdomini, in sù i Casolari de' Tedaldini. E puivvicarono lo Studio per tutta Italia, e havuti i Dottori assai famosi in tutte le facultà delle leggi, e dell' altre scienzie, cominciarono a leggere a dì sei del mese di Novembre, gli anni di Christo MCCCXLVIII."

lation of the city consequent upon the pestilence, and to reap material advantage.[8] If such was the case, the results were disappointing, for in 1357 Villani notes that the governors of Florence complained that the expense of their university was far in excess of the profits.[9] In 1364 Charles IV granted it an imperial diploma,[10] but the years immediately following the Black Death were not favorable to institutions of learning, and the difficulties of Florence continued in spite of the emperor's favor.[11]

The decline of learning consequent upon the pestilence was a matter of deep concern to Charles, himself formerly a student at Paris, and is reiterated in the introductions of five charters which he gave to as many different universities between the years 1355 and 1369.[12] The prefaces of the imperial diplomas are much alike, each declaring that precious knowledge, which the mad rage of pestilential death has stifled throughout the wide realms of the world,

[8] F. Carabellese, *La peste del 1348 e le condizione della santà pubblica in Toscana*, Florence, 1897, p. 51: "La città rimase spopolata, la Signoria provvide per l'incremento dello studio, e sotto colore di favorire le scienze e l'onore della città, pensava al suo vantaggio materiale."

[9] Villani, *Istoria*, Book 7, ch. 90, in Muratori, Vol. 14, col. 458.

[10] A. Gherardi, *Statuti della università e studio fiorentino*, Florence, 1881, p. 139.

[11] G. Tiraboschi, *Storia della letteratura italiana*, Milan, 1822–1826, Vol. 5, Pt. 1, p. 123.

[12] The universities were: PERUGIA, in the year 1355; the text of the privilege is given by P. D. V. Bini, *Memorie istoriche della perugina università degli studj e dei suoi professori*, Perugia, 1816, p. 206; SIENA, in 1357, text printed by F. Ughelli, *Italia sacra*, Vol. 3, Venice, 1718, col. 563–565; PAVIA, 1361, text in R. Maiocchi, *Codice diplomatico dell' università di Pavia*, Pavia, 1905–1915, Vol. 1, No. 1, pp. 7–9; ORANGE, 1365, text in M. Fournier, *Les statuts et privilèges des universités françaises depuis leur fondation jusqu'à 1789*, Paris, 1890–1894, Vol. 2, No. 1543; LUCCA, 1369, text in E. Baluze, *Miscellanea novo ordine digesta*, Lucca, 1764, Vol. 4, p. 184.

is calling upon the emperor to raise her from her prostrate condition.[13] Though the similarity in wording causes Father Denifle to refer to this as the emperor's stereotyped introduction,[14] and though, as he points out, portions of the description of knowledge and her fallen estate are taken from Manfred's references to philosophy in his thirteenth-century charter to the University of Naples,[15] these repeated assertions of the disastrous effects of the pestilence upon formerly flourishing sciences and universities show what an impression had been made upon the mind of the emperor. Nor was this proemium employed in all the imperial diplomas whereby Charles sought to revive

[13] Extracts from those granted Siena and Pavia, taken respectively from Ughelli and Maiocchi as given above, will give an idea of the wording of them all, and show that even though, as here, there may be differences of phraseology, the substance is the same:

"Veneranda virtutum magistra, rectrix morum, & recta humanae conversationis imbutrix, sacrarum legum, & canonum, ac liberalium artium preciosa scientia, quam pestilentis pridem mortalitatis rabies perampla orbis climata suffocavit, ipso sui silentio ad nos clamat, & invocat tacite nomen nostrum, ut ad relevandum ipsius prostratae lapsum Imperialis ei dexteram potentiae porrigamus." So, at the supplication of certain men whom he names, "studium, quod ibidem hactenus viguisse, sed his temporibus permissu Dei, aequaliter obscuratum esse dignoscitur, in lucem decernimus erigere redivivam . . ." etc. From the charter granted Siena, 1357.

"Rectrix humani generis, virtutum celestium imitatrix [or mutatrix], praeclara scientia, cuius elata membra odiosa pestilentie rabies per mundi climata impia voragine jam assumpsit, tanto gemitu ad nos clamare [or clamore] confiditur, jam suis laribus a penosis habitaculis viduata, ut sibi dignemur [or dignemini] imperialis celsitudinis presidio subvenire, si quando provisione cesarea, cuius interest universo mundo consulere, in pristino statu florere valeat et crescentem mundum sustentare et denuo fecunda quidem [or quadam] aspergine germinis redimiri." From the charter granted Pavia, 1361. The words in brackets are variations occurring in MSS other than that used by Maiocchi.

[14] Denifle, *Entstehung der Univ.*, p. 550.

[15] Denifle, pp. 457–458.

stricken universities, as is evident in his privileges to Arezzo in 1355,[16] and to Florence in 1364.[17]

A strong statement of the dire consequences of plague, as well as another illustration of the impression made upon the rulers of the people by its ravages, is found in the mandate of the duke of Anjou, governor of Languedoc, in 1376, ordering that the Montpellier faculty of medicine be given every year for dissection the corpse of an executed criminal:

> Since from the weakness of human nature and the bad mode of living of individuals it often happens that fevers, infirmities, imposthumes, and various other kinds of illness result, from these, unless suitable remedies be supplied through masters, doctors of physical science or of medicine, many persons suffering those infirmities, and almost the greater part, incur the danger of death. And thus the population, which is scanty enough owing to mortalities and epidemics that have flourished in these parts, and also to wars wherein many have perished, may be diminished to the greatest possible extent, and the world brought to nothing, . . .[18]

In documents relating to the three colleges of Cambridge and the two of Oxford whose origin may be traced to the plague, allusions are made to the depletion of the learned classes and to the fact that learning is beginning to fail. In January, 1350, Trinity Hall, at Cambridge, was founded by Bishop Bateman of Norwich, whose brother had died of the Black Death;[19] and in a deed of 6 February, 1350, giving certain revenues to the new college, he indicates that his object in establishing it was to renew the supply

[16] Quoted by Denifle, *Entstehung der Univ.*, p. 427: "et in eadem civitate longo tempore studium viguerit juxta imperialia privilegia, que propter civilium guerrarum dicuntur deperdita."

[17] See above, p. 149.

[18] See above, pp. 111-112.

[19] *Dictionary of National Biography*, Vol. 3, pp. 395-397.

of clergy, which had been so greatly decreased by the pestilence.[20] A year or two later this same Bishop Bateman completed the foundation of Gonville Hall, begun by Edmund Gonville in 1348, but unfinished at the time of his death in 1351.[21]

The College of Corpus Christi was established in 1352, under rather unusual circumstances. The increase in fees for celebrating mass that was the result of the mortality among the clergy told heavily on the gilds, and two of those at Cambridge, Corpus Christi and the Blessed Virgin, which had recently been united, conceived the idea of endowing a college whose students would be required to say masses for departed members of the two gilds.[22] Letters patent were obtained from the king in 1352 by the duke of Lancaster, alderman of Corpus Christi, and the college was founded with one master and two fellows. However, according to Cooper, it received five other benefactions in the course of the fourteenth century.[23]

Grief for the sad state of learning and the loss of learned men in consequence of the pestilence, and a desire to assist in trying to find a remedy, are the reasons given for her efforts in behalf of University Hall at Cambridge by a granddaughter of Edward I, Elizabeth de Burgh, Countess of Clare. In her will of 1355 she left bequests to the college, to which she had for some years been making gifts, and its name was changed to Clare Hall in conformity

[20] C. H. Cooper, *Memorials of Cambridge*, Cambridge, 1860, Vol. 1, p. 114.

[21] *Documents Relating to the University and Colleges of Cambridge*, London, 1852, Vol. 2, pp. 366–368.

[22] *Documents*, II, 445–446; Bass Mullinger, *The University of Cambridge from the Earliest Times to the Royal Injunctions of 1535*, Cambridge, 1873, pp. 247–249; Cooper, *Mem. Camb.*, pp. 143–148.

[23] Cooper, *Mem. Camb.*, pp. 143–148; *Documents*, Vol. 2, pp. 19–30.

with a desire expressed by her in statutes she gave it in 1359. In the preamble of these statutes she declares that a knowledge of letters is of no small advantage in every walk of life, and that, though it may be acquired in various ways, it is best obtained in a university, which then sends forth trained men for the service of God and the state. This kind of knowledge is beginning lamentably to fail, in consequence of a great number of men having been taken away by the fangs of pestilence, and the countess wishes to do what she can to restore it by increasing the resources of the college.[24]

While these three colleges at Cambridge received their endowments within ten years of the Black Death, it was not till 1362 that the former of the two at Oxford that owe their existence to the plague was established. It was then that Simon Islip, archbishop of Canterbury, who twice made charges of unbridled avarice and neglect of duty against the clergy who survived the pestilence,[25] taking practical steps to remedy the situation, received from Edward III permission to found Canterbury College at Oxford. In the royal license the king expresses his desire for a wholesome increase of clergy by means of the spread of sound teaching, which now through the present epidemic is known to have very greatly failed.[26] The archbishop's own feelings on the subject appear in a document of the

[24] *Documents*, Vol. 2, p. 117 ff.

[25] See above, p. 138.

[26] *Historical Manuscripts Commission*, Appendix to fifth report, 1874, "Dean and Chapter of Canterbury," C. T. Martin (ed.), p. 450. License granted 1362, to found college in Oxford, to Archbishop Simon Islip by Edward III; it contains the statement: "Desiderantes incrementum salubre cleri nostri, propter multiplicationem doctrine salutaris, que jam per presentem epidemiam noscitur plurimum deficisse."

following year in which, in confirming a gift made the new college by William de Islip, he writes:

> Since through wisdom, thus not acquired without sweat and labor, kingdoms are ruled and the church militant develops in accordance with justice and spreads her tents; we Simon, &c., frequently pondering upon these matters lying close to our heart, and considering that especially those truly learned and skilled in every branch of knowledge have failed to a very great extent in the epidemics, and that on account of lack of opportunity very few are pressing forward at present for the study of letters. . . .[27]

More than a decade later William of Wykeham founded New College at Oxford, for the purpose, according to the statutes, of increasing the supply of clergy, thinned by "pestilence, wars, and other miseries of the world."[28]

Declarations of the damage done to learning by the pestilence occur not only in privileges and statutes conferred upon the universities by lords spiritual and temporal, but in utterances of the universities themselves. About the year 1350 the chancellor and scholars of Oxford University made petition to the king, "showing that the university is ruined and enfeebled by the pestilence and other causes, so that their estate can hardly be maintained or protected."[29] Mute witnesses of the same facts were books and other pledges given there by the students before the pestilence; these, in 1411, were declared to be so injured by the passage of time that unless they were soon sold they would be entirely worthless.[30] Papal aid in

[27] *Hist. MSS Comm.*, 1874, p. 450, as cited in note 26 above.

[28] H. Rashdall, "New College," in *The Colleges of Oxford*, A. Clark (ed.), London, 1891, p. 151.

[29] *Collectanea*, 3d series, London, 1896, M. Burrows (ed.), Part 2, Lucy T. Smith (ed.), p. 137. This is given in abstract, translated, taken from Anc. Pet., File 132, No. 6593.

[30] H. Anstey, *Munimenta academica*, London, 1868, Vol. 1, p. 255.

recovering their books, which had been similarly pawned, was one of the objects sought by the students of Avignon in a roll of 1361 addressed to Pope Innocent VI, beginning:

Most holy Father, at a time when the university body of your *studium* at Avignon is deprived of all lectures, since the whole number has been left desolate by the death from pestilence of doctors, licentiates, bachelors, and students, some also of the survivors of the same *studium*, who have spent many sleepless nights in the acquisition of holy canonical knowledge, are unable because of the ravages of wars, as is also the case with some of them and of the others because they are weighed down by contests over benefices due them and by the burden of poverty, to be of service to themselves and others, to recover their books, or to be promoted to the degrees which they deserve.[31]

Complaints, without such specific enumeration of causes, of the decline of learning, of decrease of students and men of learning and of poverty and abuses in the universities are frequent in the two decades following the Black Death. They were made by the universities, by outside authorities in documents relating to universities, and in writings of prominent men of the time. Where French universities were concerned, the Hundred Years' War must be borne in mind as an important cause of decline, especially in complaints like those of Toulouse and Orleans below. But even in these two instances pestilence must have been a contributory factor, while it was perhaps of greater importance than the war when the difficulties were attributed to abuses within the universities.

Between 1360 and 1365 Toulouse, Montpellier, and Orleans voiced distress. In the former year a supplication to Innocent VI from Toulouse requested investigation of

[31] M. Fournier, *Les statuts et privilèges des universités françaises*, Vol. 2, No. 1248.

the fact that, owing to the evils of the times, endowments which had formerly provided for the wants of poor students had been alienated and wrongfully seized, so that the students received little or nothing of what was due them.[32] In 1365 Orleans expressed thanks to Urban V for having rendered assistance that had made it possible for faculty and students to remain at the university. Otherwise they would have been driven to leave by poverty, due partly to their own lack of resources, partly to the small number of ecclesiastical and worldly contributions, since relatives and friends who had formerly supported them had by wars been reduced to want.[33] Three years earlier a roll of Montpellier to the same pope had set forth in striking fashion the straits to which she was reduced, and made a plea for the restoration of knowledge, which was threatened with destruction; "at present the said *studium* is destitute of lecturers and listeners, because in it, where formerly a thousand students used to dwell, scarcely two hundred are to be found today." So divine and papal aid are sought, that the *studium* may be reformed "and knowledge, the loss of which is to be feared, may be restored."[34] Temporal assistance also was invoked in a supplication of the latter half of the fourteenth century, which the students requested the grand council to transmit to the consuls of Montpellier. In it, after dwelling upon the former world-wide fame and prestige of the university, they lament that now, alas, scarce a vestige of these remain. In giving bad management, negligence, and unbridled cupidity on the part of the masters regent as the "cause of this decline and most recent desolation," the

[32] Fournier, Vol. 1, No. 640.

[33] Fournier, Vol. 3, No. 1886.

[34] *Cartulaire de l'Université de Montpellier,* Montpellier 1890, Vol. 1, No. 119.

students speak as if it were the latest of several periods of distress.[35]

These evils gave rise to serious concern on the part of popes and other authorities, and there is evidence of earnest efforts to remedy them. Before 1348 the chartulary of the University of Montpellier and Fournier's collection of documents concerning it contain no mention of abuses. but afterwards complaints or decrees concerning them occur in 1352, 1355, 1362, 1364, 1373, 1382, and 1384, coming from, respectively, King John, a proctor, three successive popes, and a cardinal.[36] For Toulouse, Urban V in 1365 appointed a commission to make inquiry into the disorders and abuses of the *studium*.[37]

There are other official statements of decline regarding these and other universities. Paris seems to have been strong enough to pass through the plague without expression of injury, but there are some indications to show that she felt it. In 1351 Clement granted that Bernard de Pacterius of the Friars minor be allowed to lecture as bachelor in the *studium* at Paris, "especially since there is there so great a scarcity of bachelors of that order that there are not enough to perform the scholastic and disputatious functions of masters."[38] A year later he granted to the chancellor of Paris a like permission on behalf of William Romanus of the Friars Preachers, for the even more striking reason that there was so great a scarcity of bachelors of his order in the university that often there was no one else who was actually lecturing.[39] In June and July of the same year similar permits were issued, though

[35] Fournier, Vol. 2, No. 1060.
[36] Fournier, Vol. 2, Nos. 980, 983, 989, 993, 1018, 1028, 1034.
[37] Fournier, Vol. 1, No. 667.
[38] *Cart. Univ. Paris.*, Vol. 3, No. 1198.
[39] *Cart. Univ. Paris.*, Vol. 3, No. 1206.

with no reason assigned, to two Augustinians.[40] A few years later Edward III appealed to the bishops for help in restoring Oxford, which students had forsaken and which had become "like a worthless fig tree without fruit."[41] At about the same time another monarch, Charles IV, in his privilege of 1355 to the *studium* of Arezzo, attempted to revive that formerly flourishing institution. The attempt was not successful, and from 1373 to 1456 all traces of the school disappeared.[42]

By three successive popes the failure of learning, the decrease in the number of students, and the lack of learned men are given in the bulls of foundation of colleges as the reasons for their endowments. In that of the College of St. Martial at Toulouse, in 1359, Innocent VI provides for ten students in civil and ten in canon law, that these studies may not cease to flourish, "which we with grief perceive to be failing, whether by reason of sluggishness of wits, or of human sloth, or of the evils of the times."[43] Ten years afterward Urban V, in establishing the College of Douze-Médecins at Montpellier, says that he does so because in that university, in which has flourished for so long a time a *studium generale*, producing preëminent men imbued with the teachings of health-giving knowledge, there are at present few students in the faculty of medicine, and that he wishes to increase their number.[44] Cardinal Egi-

[40] *Cart. Univ. Paris.*, Vol. 3, Nos. 1207 and 1208.

[41] F. A. Gasquet, *The Great Pestilence*, London, 1893. p. 210, quoting from *Reg. Trileck*, fol. 163.

[42] Denifle, *Entsteh. d. Univ.*, pp. 427–428. The emperor, in the privilege, attributes the decline to civil wars.

[43] Fournier, Vol. 1, No. 617.

[44] *Cart. Univ. Mont.*, Vol. 1, No. 150; Fournier, Vol. 2, No. 1010. Petrarch, also, in "Epistolarum de rebus senilibus," lib. 10, ep. 2, p. 868 of *Opera quae extant omnia*, Basel, 1581, wrote in 1367 of the changes which had come over Montpellier since his youth,

dius Albornoz, who died in 1367, left instructions for the founding of a college at Bologna for poor Spanish students.[45] The college was already functioning when, at the request of the cardinal's testamentary executors, Gregory XI formally established it by a bull of 21 September, 1371, as the College Albornoz. In this bull he gives as the cardinal's reason for the foundation the needs of some churches in parts of Spain, which were suffering every day serious harm because of the falling off of men learned in spiritual and temporal subjects; and also the poverty of those regions, which prevented gifted students from being able to study.[46]

In the annals of the city of Perugia for 1385, Bini finds it recorded that the *studium* there had for many years, through neglect, been reduced to nothing.[47] Famous scholars add the weight of their evidence to the testimony of kings, popes, and city fathers regarding the decay of learning and universities within the twenty years following the Black Death. Petrarch, in his letter, cited above, to the archbishop of Genoa, after telling him of the distressing change at Montpellier, goes on to speak of Bologna:

> Thence we journeyed to Bologna, than which I think there used to be nothing more joyous, nothing freer, in the world. Surely you remember what a crowd of students was there, and what their order was? what their vigilance? what that of their instructors? You might consider actually divine those ancient

"florentissimum tunc oppidum, jurisque ad studium delati," and asks concerning it at the present time, "Quae scholarium turba? quae copia magistrorum?"

[45] E. Baluze, *Vitae paparum avenionensium*, Paris, 1914–1922, Vol. 1, p. 364.

[46] *Chartularium studii Bononiensis*, Bologna, 1909–1923, Vol. 2, pp. 261–262.

[47] P. D. V. Bini, *Memorie istoriche della Perugina Università*, Perugia, 1816, pp. 52–53, quoting Ann. Dec. an. 1385, fol. 87.

jurisconsults, of whom today scarcely a single one is there; but in the place of so many and such great geniuses, a universal ignorance has seized that city . . .[48]

The decrease in number of students at Oxford is shown to have been very marked by the estimates of three scholars who were connected with it. In 1357 Richard Fitzralph, who had been chancellor of Oxford both before and after the pestilence, conjectured that it might have been because of the pernicious activities of the friars that "whereas in the *studium* of Oxford formerly in my time there were 30,000 students, today there are not 6,000."[49] Wyclif makes the loss even greater, from 60,000 to less than 3,000, and apparently is inclined to place the blame upon changes in worldly possessions of the church;[50] but Thomas Gascoigne, about a century later, supports Fitzralph's figures, saying that, when he was chancellor at Oxford, he had seen in the rolls of the ancient chancellors that the number of scholars there before the pestilence was 30,000.[51] These figures are usually considered decided exaggerations, but they show that the loss of students was very great; indeed Fitzralph in the same work, written in 1357, observed that the attendance was never more than a fifth of what it had been.

It is well to note the reasons for the evil estate of learning other than pestilence and, in France, the Hundred

[48] Petrarch, *Ep. senil.*, lib. 10, ep. 2.

[49] Richard Fitzralph, "Defensorum curatorum contra eos qui privilegiatos se dicunt," 1357, in E. Brown, *Fasciculus rerum expetendarum et fugiendarum,* London, 1690, Vol. 2, pp. 473–474.

[50] John Wyclif, *De ecclesia,* J. Loserth (ed.), London, 1886, p. 374. "Tunc enim prosperaretur regnum in corporalibus et spiritualibus et cresceret universitas nostra tam numero quam virtute; in cuius signum ante catheclismum appropriacionis huiusmodi erant in Oxonia sexaginta milia scolarium, ubi hodie non sunt tria."

[51] See above, pp. 132–133, note 42.

Years' War. Petrarch, as does Tiraboschi in quoting him, blames wars for the plight of Bologna,[52] and we have already seen war mentioned with plague in the complaints of several universities. Fitzralph attributes the decline of Oxford to the friars; Wyclif to the grasping spirit of the clergy, though his "catheclismum appropriacionis" seems to date from about the time of the pestilence; Gascoigne attributes it to the Black Death. Mullinger, in his history of Cambridge, suggests interest in polemics and civil law as being to some extent an answer to his question:

> How is it that from the middle of the fourteenth century up to the revival of classical learning [he is speaking of England], the very period wherein the munificence of royal and noble founders is most conspicuous in connection with our university history, such a lull comes over the mental life of Oxford and Cambridge, and so few names of eminence, Wyclif and Reginald Pecock being the most notable exceptions, invite our attention? From the death of Bradwardine[53] to the first battle of St. Albans, more than three quarters of a century intervene, during which no adequate external cause of distraction appears which may be supposed to account for the comparative inertness of the universities.[54]

He has elsewhere commented upon the Black Death as the cause of the foundation of the three colleges at Cambridge to which reference has been made.[55] And among other authorities on university history, Denifle,[56] Wood,[57] and Tiraboschi[58] emphasize the disastrous immediate effects

[52] Tiraboschi, V, pt. 1, 85–88.

[53] Died of the Black Death in 1349. See above, p. 125.

[54] Mullinger, *University of Cambridge*, II, 208.

[55] Mullinger, I, 242.

[56] Denifle, *Entsteh. d. Univ.*, pp. 274, 361, 550.

[57] Anthony Wood, *The History and Antiquities of the University of Oxford*, Oxford, 1792, Vol. 1, pp. 449, 453–454, 476.

[58] Tiraboschi, V, pt. 1, pp. 70–139.

of this pestilence upon the universities. Due weight must be given to wars, disputes, and polemics, but they had occurred before 1348, and in the late fourteenth and fifteenth centuries they were still raging: nevertheless, it is only in the generation following the Black Death that testimony comes from all sides of the decline of universities, the failure of men of learning, and the threatened extinction of knowledge. Nor do these declarations fail in many instances to give the plague as cause, a charge so often repeated that Rashdall suggests in the case of the statutes of New College, that it has become merely a rhetorical flourish.[59]

There are other indications of decline, and of recovery dating from the last quarter of the century, that are scattered and indefinite, but are of some complementary value. The first is the number of universities which were existing in the early forties, but which had disappeared by the middle of the century. Their cessation is usually marked by no act or comment, but by a disappearance of further documentary material about them. Such was the case with Grenoble,[60] Vercelli,[61] Reggio, and Naples; Verona and Friuli, founded about this period, apparently never opened. Unsuccessful efforts were made at about the same time to revive the University of Rome, never very prosperous and then moribund; and Arezzo, despite imperial privilege, practically passed out of existence from 1373 till 1456.[62]

[59] Rashdall, in *Colleges of Oxford*, p. 151.

[60] Fournier, Vol. 2, Nos. 1546–1554, gives nine documents relating to Grenoble, their dates falling between 1339 and 1345.

[61] H. Rashdall, *Universities of Europe in the Middle Ages*, Oxford, 1895, Vol. 2, Pt. 1, p. 27.

[62] Tiraboschi, V, pt. 1, 102–103 for the University of Naples; pp. 125–138 for Reggio, Verona, Friuli, Rome (not the Court of Rome), and Arezzo.

Siena was more fortunate, for, reopened by imperial decree in 1357, she flourished during the rest of the century.[63] Pavia suffered also,[64] and the effect on Perugia has been mentioned.[65] As regards Pisa, Fabroni, after enumerating famine, wars, and taxes as causes of her fall from her eminence of 1338, concludes with the very fearful pestilence of 1348, which is remembered by all writers with tears, and which, in addition to the other misfortunes mentioned, made it seem that the university was tottering on the verge of destruction.[66] His assertion is supported by two extracts cited by Tiraboschi from the archives of the city, both decrees of the citizens. The first, of 1359, ordains that no more professors be brought there to teach civil and canon law, and that those who are already there doing so must be licensed; according to the second, of 1362, Master Francis of Cremona is bound to teach alone, through the following year, grammar, logic, and physics to all desiring to take them in the schools of the city of Pisa.[67] And not only did universities founded before the Black Death suffer difficulties and dissolution, but those which were established between 1348 and 1370 attained success either not at all or after the latter date: Prague, for example, Florence, Piacenza, Geneva, Cracow, Vienna, and Fünfkirchen.[68]

A second indirect sort of evidence may be obtained from a comparison of attendance given at different periods of the

[63] Denifle, *Entsteh. d. Univ.*, p. 447.

[64] Z. Volta, "Dei gradi academici conferiti nello 'Studio Generale' di Pavia sotto il dominio Visconteo," in *Archivio storico lombardo*, Ser. 2 (1890), p. 517.

[65] See above, p. 159.

[66] A. Fabroni, *Historiae Academiae Pisanae*, Pisa, 1791, Vol. 1, p. 70.

[67] Tiraboschi, V, pt. 1, 107–108. He gives the wording of the second decree, but only the substance of the first.

[68] Denifle, *Entsteh, d. Univ.* 1885.

century in rolls sent the popes. Until almost the end of the century such rolls are infrequent and inaccessible, and where they exist earlier they contain the names only of professors (doctors, masters), and masters, licentiates, and bachelors. Furthermore, it is difficult to group them in any satisfactory way, since the titles do not always have the same meaning. Master, for instance, sometimes denotes a professor, sometimes a master of arts who is studying as a bachelor in one of the higher faculties. Hence the tables on page 165, taken from the rolls of those universities which offer enough data for comparison, are of value chiefly for their suggestiveness. The first furnishes statistics of members of faculties and advanced students both for the generation of decadence following the Black Death, and for the succeeding period of revival.[69]

Thus there was a total of fourteen professors and one hundred thirteen licentiates and bachelors at the four universities when the earlier rolls were drawn up, compared with fifty and eleven hundred twenty-one respectively at the time of the later. It will be remembered that Toulouse, Avignon, and Orleans had all made urgent representations of distress to the popes between 1360 and 1365. From Montpellier, who also declared her desperate plight at this time, there are two rolls of 1362. One, of 24 November, is a roll of the university, and contains 44 names, chiefly bachelors; the other, two days later, of the faculty of medicine, consists of 28 names, almost all bachelors of medicine.[70]

[69] Fournier, for the different universities as follows: Angers, Vol. 3, supplement, Nos. 1895, 1897; Avignon, Vol. 2, Nos. 1247, 1248, 1269; Orleans, Vol. 3, supplement, Nos. 1881, 1886, 1888; Toulouse, Vol. 1, Nos. 656, 697.

[70] *Cart. Univ. Mont.*, Vol. 1, Nos. 119 and 120.

1349–1365

	Professors (Masters, Doctors)	*Masters of Arts, Licentiates, Bachelors*
Angers (1363)	1	45
Avignon (1353)[a]	2	
Avignon (1361)	1	8
Orleans (1349)[a]	8	38
Orleans (1365)[a]	3	40
Toulouse (1362)	6	21
Total (averaging when more than one roll for each university)	14	113

1370–1394

Angers (1378)	13	438[b]
Avignon (1393)	11	161
Orleans (1378)	5	230
Toulouse (1378)	21	292
Total	50	1121

[a] In addition, there are given at Avignon in 1353 two clerks; at Orleans, both in 1349 and 1365, ten other names, without indication of rank. None of them, however, seems to belong in the classifications given.

[b] Including seventy-seven absent.

Rolls of the faculties of the University of Paris of 1349, 1362, and 1387,[71] because of arrangement different from those above, must be given separately. They seem to

[71] *Chart. Univ. Paris.*, Vol. 2, Nos. 1162–1165, for rolls of 1349; Vol. 3, Nos. 1262–1265, for rolls of 1362; Vol. 3, Nos. 1537–1541, for rolls of 1387.

show that the professors of the three higher faculties, and the graduate students of the faculty of arts, were not so much affected at the time of the outbreak of the pestilence as was the succeeding generation, particularly since the latter suffered also from the attack of plague of 1360. It is unfortunate that there exists no similar roll from the earlier part of the century, though it is true that the roll of 1349 may have been made before the Black Death had taken its toll in Paris:

	1349	1362	1387
Faculty of Theology..........	32	25	15
Faculty of Law..............	17	11	16
Faculty of Medicine..........	46	26	20
Faculty of Arts..............	501	446	410

Owing to the fact that the rolls had to be taken when they could be found, it was not possible to distinguish between the consequences of the Black Death and those of the pestilence of 1360. In fact, since the full consequences of such a visitation cannot satisfactorily be estimated within less than a generation, it is necessary to go past the date of the latter at times throughout this whole discussion. But in a table of comparative number and mortality of professors at Padua, based on Gloria's *Monumenti della Università di Padova,* as well as in the comparison of the number of documents appearing at different times in university chartularies, the material can be arranged so as to view separately the results of the Black Death.

Padua, with Paris, is distinctive in apparently having made no sign of hurt during the pestilence or afterward. Hence it is interesting to find that among the professors carefully listed by Gloria the rate of mortality for the period of the Black Death is 16⅔ per cent, three times as high as for periods of the same length shortly before and after it.[72]

	1344–46 *Professors*		1347–49 *Professors*		1357–59 *Professors*	
Total number	13	3[a]	21	3[a]	35	3[a]
Number died	1	0	3	1	1	0

[a] Given as professors by some authorities, but considered doubtful by Gloria.

Nevertheless Padua continued to function normally, holding doctoral examinations and conferring degrees while the plague was at its worst.[73]

The number of university documents for the years following the Black Death is as a rule noticeably smaller than at other times. For instance, from the University of Bo-

[72] A. Gloria, *Monumenti della Università di Padova,* Padua, 1888, Vol. 1; and, by the same author, "Studi editi dalla Università di Padova," Vol. 1, published in *Istituto Veneto Memorie,* Vol. 22[3]. Page references are hardly practicable, since the names and facts used to compile the table were gathered from all parts of the two volumes, though especially from the volume of *Monumenti.*

[73] A. Gloria, *Monumenti;* I, 19, comments on this in an interesting passage: "Not even the earthquake and the plague of 1348 occasioned suspension of the university, since we read that Professor Dino da Urbino was in the episcopal curia 11 February, 5 April of that year, and in company with the scholar Vitaliano Vitaliani 30 April; that Giordano de Halpeze obtained the doctorate 3 March; that it was obtained 6 April by Aichino Orsi Carnelli di Chioggia, dead a little afterwards of that pestilence; and that the monk Vitale da Lodi also obtained the doctorate on the tenth of April."

logna we have jurist statutes of 1317–1347,[74] then next those of 1432;[75] while the only two fourteenth-century rolls for the faculties of law and arts are for the academic years 1384–1385, and 1388–1389.[76] A count of the documents given in Fournier concerning the five universities whose records extend through the period[77] gives the following results:

NUMBER OF DOCUMENTS

	1335–47	1348–60	1361–73
Angers	7	4	13
Avignon	1	3	9
Montpellier	42	18	41
Orleans	46	14	33
Toulouse	38	23[a]	64
	134	62	160

[a] Only 4 to 1358.

The efforts to counteract the disastrous consequences of pestilence upon the universities took the form chiefly of founding bursarships and colleges; more of the latter were established in the last half of the fourteenth century than

[74] H. Denifle, "Die Statuten der Juristen-Universität Bologna," in *Archiv für Litteratur und Kirchengeschichte des Mittelalters*, III, 201–203. They were found by Denifle after a long search in the Capitelsbibliothek at Pressburg, in a manuscript the apparatus, painted initials, etc., of which agree with the juristic writings of Bologna of the mid-fourteenth century.

[75] C. Malagola, *Statuti delle Università e dei Collegi dello Studio Bolognese*, Bologna, 1888, p. 3.

[76] U. Dallari, *Rotuli dei lettori, legisti, e artisti dello Studio Bolognese dal 1384 al 1799*, Bologna, 1888, I, 6.

[77] Fournier, *Statuts et privilèges*. The documents are arranged chronologically under the different universities.

in any other epoch of medieval history.[78] Paris again is an exception, as are Oxford and Cambridge to a lesser extent.

TABLE OF COLLEGES FOUNDED AT ELEVEN UNIVERSITIES BEFORE 1400

	Till 1300	1301–1348	1349–1400
Angers	0	0	1
Avignon	0	0	2
Bologna	1	1	3
Cambridge	1	4	3
Orleans	0	0	0
Oxford	3	3	3
Padua	0	0	6 or 7
Perugia	Not founded till 1307	0	2
Toulouse	2	2	3
	7	10	23
Paris	18	23	15
Total	25	33	38

The founding of universities in the second half of the century shows a 50 per cent increase over that of the first: from 1300 to 1348 twelve opened their doors to students; eighteen, from 1348 to 1400.[79] There is little indication of the importance of the Black Death as a factor here, except

[78] The data for compiling this table were gathered from the various sources cited above, especially chartularies and other documentary collections. An exception is found in the colleges of Paris, which are taken from the list given by Rashdall, Vol. 1, pp. 514–517. A work not cited before, which is here used in connection with the colleges of Padua, is J. Facciolati, *Fasti Gymnasii Patavini*, Padua, 1757, pp. 120–129. The table is complete only to the extent to which these sources were exhaustive.

[79] Denifle, list of universities cited above, note 1, p. 146.

in contributing to the failure of some and delaying the success of others. But it seems a fair presumption, especially in view of the granting to many of them of the hitherto jealously limited theological faculties, that the same crying need of clergy voiced in the founding of colleges played its part morally and economically in the establishment of universities where formerly none had been.[80]

This trend toward standardization dating from the middle of the fourteenth century is shown in various ways. Not only was the study of theology allowed in new universities, but it was introduced into many from which it had always been excluded.[81] The status of *studium generale,* or a place where all sorts of learning could be taught, according to Denifle, was more and more sought and obtained after the middle of the fourteenth century,[82] and it was then that there began to grow up the idea of the union of teachers and students in a corporation, with the privilege of university right. At this time, too, the students acquired more and better acknowledged rights in the management of university affairs, which is considered by Denifle a definite advance. The earlier canonists, he says, John Teutonicus and John Andreas, like Accursius and Odofredus, maintained that not the scholars but only the professors had the right of entering into contracts. But whoever recognized that the students might elect their rector had to acknowledge also that they might enter into contracts, and it was toward the middle of the fourteenth

[80] F. Ehrle, "Die Ehrentitel der scholastischen Lehrer des Mittelalters," in *Bayerische Akademie der Wissenschaften Sitzungsberichte,* Munich, 1919, 9. Abhandlung, p. 25.

[81] J. Facciolati, *De Gymnasio patavino syntagmata XII ex ejusdem gymnasii fastis excerpta,* Padua, 1752, pp. 78–79; Tiraboschi, *Stor. d. lett. ital.,* V, 239; Denifle, *Entsteh. d. Univ.*

[82] Denifle, p. 27.

century[83] that the former right, though not yet the latter, was declared by Bartolus and Baldus to belong to the students.[84] Further, the jurist statutes of Montpellier of 1339 give no information as to whether the rector must be a student, as he had been at Bologna since the middle of the thirteenth century; but those of 1351 expressly state that he must be.[85]

Associated, probably, with the increased prominence of students in the management of university affairs is the active part taken by the universities of Oxford in England and Paris in France in political affairs. The former, after throwing off episcopal control in 1350, became a center of free thought[86] and of Wyclifism; the latter, in the late fourteenth and early fifteenth centuries, was a bold and recognized leader in efforts to better the government of France and to infuse spirit and order into the conduct of the war against the English.[87] A similar catastrophe befell them both. At Oxford suppression of free thought in 1411, and at Paris a decade later acceptance of English rule, closed for each the period of medieval activity.

A further manifestation of the effects of the Black Death, appearing in much of the writing and art of this and following generations, is the somber shadow it cast upon men's minds:[88] a reflection of some importance in

[83] Denifle, p. 23.

[84] Denifle, pp. 172–174.

[85] Denifle, p. 186.

[86] Rashdall, *Univ. of Europe,* Vol. 2, Pt. 2, pp. 426–435.

[87] A. M. Campbell, *Evidences of a Spirit of Nationalism in the University of Paris in the Late Middle Ages,* master's thesis at Columbia University, 1924.

[88] For an interesting discussion of this, especially in connection with the sixteenth and seventeenth centuries, see Lynn Thorndike, "The Blight of Pestilence on Modern Civilization," *American Historical Review,* April, 1927, pp. 472–473. Of interest also in the statement, from *Histoire littéraire de la France,* Vol. 24, p. 274:

view of subsequent developments in mysticism, reform, and puritanism. In the testaments and other instruments endowing colleges during the last half of the century, the idea of death is unwontedly prominent. Before then the reasons given for such benefactions occasionally refer to poor health and the approach of death, but afterwards not only does it become the rule rather than the exception, but there occurs a new departure in philosophic and somewhat fatalistic passages concerning death. For instance, once more basing our estimates principally upon Fournier, whose collection of documents is unrivaled for material bearing upon medieval colleges, we find that of the colleges which he records as being founded before 1348, the first is established at Toulouse in 1243, by will of Vidal Gautier, for the honor of God and the Virgin and for the salvation of his own and his ancestors' souls. The donor has stated that he has a serious disease of which he is dying, but he makes no observations on death.[89] In 1263 precisely the same reasons are given by James I of Aragon for the founding of the theological college of Valmagne, with no mention of death here or in Clement IV's bull of confirmation two years later.[90] The same is true of the bishop of Bayeux's endowment at Paris, in 1309, of a college for scholars of Le Mans and Angers: he does it, he says, because of the advantage of good scholars to church, state, and people.[91] In 1319 William de Montlezun began at

"Nous rencontrerons à tout moment les témoignages de la profonde impression de ce fléau sur les esprits. Le jurisconsulte Henri Bohic, dans son commentaire sur les decrétales, dit qu'il se hâte, pour n'être point prévenu par le mort."

[89] Fournier, Vol. 1, No. 517.

[90] Fournier, Vol. 2, Nos. 892, 893; *Cart. Univ. Mont.*, Vol. 1, Nos. 10, 11.

[91] Fournier, Vol. 1, No. 372.

Toulouse the foundation of Montlezun College, for the honor of God and His Mother the Virgin, of all the angels and saints and the whole heavenly court, for the acquisition of the pearl of Catholic knowledge.[92] The first observation on death comes from Galbert, archbishop of Arles, when endowing at the same university in 1341 the College of Narbonne, in preparation for the day of death, which comes to all without distinction.[93]

Noticeably different is the tenor of some of the succeeding foundations; the note is struck in Innocent VI's bull of 1353 authorizing Bernard Trigard, bishop of Pézenas, to found at Montpellier the college of Pézenas, inasmuch as he, thinking of his own salvation, desired by a happy exchange to commute his earthly possessions into heavenly ones, and that which was transitory into that which was eternal.[94] Five years later, the bishop, in his will in favor of the college, observed that the joining of the human condition to the soul subjected the latter, from the beginning, to the hazards of inevitable death.[95] The same pope, who had succeeded within a few years of the Black Death to the governance of plague-stricken Christendom, in his own foundation, in 1359, of the College of Saint Martial at Toulouse, voiced at some length his philosophy of death.[96] After dwelling upon God's unequal distribution of things in this life, he emphasizes the absolute impartiality of His application to all of the law of death:

> For the lowest dies with the most exalted; with the common folk perish the nobility; the learned equally with the unlearned.

[92] Fournier, Vol. 1, No. 549.
[93] Fournier, Vol. 1, No. 595.
[94] *Cart. Univ. Mont.*, Vol. 1, No. 105.
[95] Fournier, Vol. 2, No. 985.
[96] Fournier, Vol. 1, No. 617.

Nor is there object in our extolling the vain glory of this world with empty titles, when the day of death is set for us at birth, and through still and hidden courses of this sort the mortal part of our life is hurried along by the fleeting transience of the times . . . if, indeed, that can be called life, wherein we continually are dying and gliding away like water, and which flees as the shadow, and never remains the same . . .

A popular quotation in foundations of the period is St. Augustine's "nothing is more certain than death, nothing less so than the hour of its coming," which occurs in three documents of the kind between 1361 and 1365, one of 1371 quoted below, and another of 1385.[97] They are the establishments, respectively, of Fougères College at Angers University by William Georges; of Maguelone College at Toulouse by Cardinal Audouin Aubert, who includes two other statements about death from St. Augustine, and one from the Psalms; of Pelégry College at Cahors by Raymond of Pelégry; of Rodez College at the same university by Bernard of Rodez, archbishop of Naples; and a will in favor of the College of Saint Catherine at Toulouse by Peter de Monteruc. In the opening paragraph of his foundation charter for Rodez College, the archbishop remarks that he has been pondering upon the facts that all men die once, that all flesh and its glory fades like the flower and returns to dust, that the life of man is only a vapor or noxious exhalation appearing for a time above the earth, and that nothing is more certain than death, nothing less so than the hour of its coming. He adds that he is not expecting death.[98]

Data about elementary schools at this time are scarce, and as they are not here of such importance as to justify

[97] Fournier, Vol. 1, Nos. 385, 659; Vol. 2, Nos. 1430, 1441; Vol. 1, No. 711.

[98] Fournier, Vol. 2, No. 1441.

the prolonged search that adequate treatment would require, the facts are those which have come to light in the course of other lines of investigation. Usually they have to do with the engagement and salaries of teachers. Carabellese cites a good many items of this sort from the archives of the senate of Lucca for the years 1344–1352; one, of 1348, deserves to be quoted here.

On the fourteenth of August the senators had "seen and examined a certain petition of the following tenor, namely, it is set forth to you reverently on the part of Master Philip, teacher of boys, that he, who for a long time has been in the service of the citizens of Lucca in teaching their sons, on account of the calamity of a time of great ruin and decline has been reduced to want. Nevertheless he does not wish or know how to cease teaching boys, yet, because of the poverty of the citizens and of a number of the boys, he would not be able to live by the boys' sole provision." The senators pledged themselves to a subsidy of 3 lire a month, "since, on account of the pestilence that occurred in this year, all the teachers of the boys are dead, and lest the boys in consequence roam about idly, and by their long vacation from teachers become ignorant of letters, they had better come under the teacher's rod for the virtue of learning." [99]

Another Italian schoolmaster, Francesco Agezzi of Vercelli, testifies that after the Black Death he had scarcely forty pupils, while before he used to have two hundred;[100] and Lechner quotes from Sanudo the fact that an inscription over the door of the charity school in Venice stated that the rector, ten teachers, and more than three hundred pupils died of the plague.[101] Also he gives statistics for

[99] Carabellese, *La peste del 1348,* pp. 53–54, note 2, citing *Anziani* 28 ac. 36. t.

[100] G. Manacorda, *Storia della scuola in Italia; il medio evo,* Milan, 1914, Vol. 1, p. 155.

[101] Karl Lechner, *Das grosse Sterben in Deutschland in den Jahren 1348 bis 1351 und die folgenden Pestepidemien bis zum Schlusse des 14. Jahrhunderts,* Innsbruck, 1884, p. 56.

salaries of teachers of Latin schools in Ragusa from 1346 to 1360: in 1346 Master Andreas received a yearly salary of 20 iperperi,[102] and might require tuition fees of his pupils; in 1347 Master Peter received 35 iperperi and tuition fees; in 1357 Don John received 40 iperperi and a set fee of two groschen monthly from each pupil who learned Donatus; in 1360 the maximum teacher's salary was set at 60 iperperi.[103] This also seems to point toward increased emolument for the professional classes as a result of the plague.

The continuer of the Chronicle of Guillaume de Nangis says that in France after the Black Death "few were found in houses, villas and castles who were able or willing to instruct boys in the rudiments of grammar."[104] In England the records of the York Archaeological Society for 1368 give the appointment of Simon of Bekyngham, chancellor of the church of the Blessed Peter at York, as schoolmaster for life instead of for the usual three-year period, on account of the scarcity of masters of arts as a result of the pestilence.[105] William of Wykeham, whose grammar school at Winchester was the beginning of the English public school system, in 1373 engaged Master Richard of Herton *(Grammaticus)* for ten years to teach such poor scholars as William presented, and none other.[106] John of Trevisa's

[102] Fourteen iperperi made a pound, according to Lechner.
[103] Lechner, pp. 75–76.
[104] H. Géraud, (ed.), Guillaume de Nangis, Chronique latine de 1113 à 1300; avec les cotinuations de cette chronique de 1300 à 1368, Paris, 1843, Vol. II, 216.
[105] A. F. Leach (ed.), *Early Yorkshire Schools,* London, 1899, p. 23, "Yorkshire Archaeological Society Records," Ser. 27, quoting Acta Cap. B. cij 86.
[106] R. Lowth, *Life of William of Wykeham,* London, 1758, pp. 359–360. This is Document VII in the appendix, E Registro Wykeham. Part. Tert. a. fol. 98.

statement about certain schoolmasters introducing into English grammar schools after the pestilence the use of English instead of French as the language of instruction, dates the beginning of the change from the "furste moreyn."[107] The Hundred Years' War suggests itself as a reason for the abandonment of French for English of at least equal importance with the thinning out by plague of teachers and officials qualified to use a language other than their own. But it is to the latter cause in great measure that the in-

[107] R. Morris and W. Skeat, *Specimens of Early English*, Oxford, 1889, pt. 2, pp 241–242; or K. Sisam, *Fourteenth Century Verse and Prose*, Oxford, 1923, p. 149. Gasquet, *Great Pestilence*, p. 234, makes use of this passage from Trevisa, giving credit to Owen and Blakeway, *History of Shrewsbury*, but with a reference to Higden's *Polychronicon*, "Rolls Series," VIII, 344. I could not locate it there, or elsewhere in the nine volumes of the *Polychronicon*, appearing, with Trevisa's translation, in the "Rolls Series." Both of the works cited above, in which the extract is reproduced, take it from a contemporary manuscript of Trevisa's translation of the *Polychronicon* different from that used by the "Rolls Series." In the portion of the passage given here, *th* is substituted for the original þ and one or two other changes of letters are made.

"Thys manere was moche y-usde to-fore the furste moreyn, & ys seththe somdel ychaunged. For Iohan Cornwal, a mayster of gramere, chayngede the lore in gramer-scole, & construccion of Freynsch in-to Englysch; & Richard Pencrych lurnede that manere techyng of hym, & other men of Pencrych; so that now, the yer of oure Lord a thousand thre hondred foure score & fyve, of the secunde kyng Richard after the conquest nyne, in al the gramere-scoles of Engelond children leveth Frensch & construeth & lurneth an Englysch, and habbeth ther-by avauntage in on syde & desavauntage yn another; here avauntage ys, that a lurneth here gramer yn lasse tyme than childern wer ywoned to do—disavauntage ys, that now childern of gramer-scole conneth no more Frensch than can here lift heele, & that ys harm for ham, & a scholle passe the se & travayle in strange londes, & in meny caas also. Also gentil men habbeth now moche yleft for to teche here childern Frensch."

creasing ignorance of Latin[108] in the schools must be attributed,[109] leading to the translation and even the production in the native languages, at this period, of works of science to be used for purposes of instruction. Such were the first and standard text in English on the astrolabe,[110] written by Geoffrey Chaucer for his young son Lewis; the fourteenth century *Tretis of Geometrie,* a translation of the second part of Robert the Englishman's *Ars metrica;*[111] and the translation into English in 1397 of the *De proprietatibus rerum* of Bartholomew the Englishman.[112] Important contributions to the French language were Nicholas Oresme's translations in 1370 and 1371, at Charles V's request, of Aristotle's *Ethics, Politics,* and *Economics;* of his own *Petit traictie de la première invention des monnoies et des causes et manières d'icelles,* which he had written in Latin; and in 1377 of Aristotle's *De caelo et de mundo.*[113] He wrote in French *Le traictie de la sphère* which, according to Cantor, from that time on became the standard of

[108] On this subject Wood, *Univ. Oxford,* I, 454, comments under the year 1352, "As for learning also it suffered such a wonderful eclipse through all parts of Europe (occasioned by the Pestilence) that the like was never before known. 'Late id temporis' (saith an author speaking of this time) 'ingens barbaries totam Europam occupavit, et sensim declinante Imperio Romano, decrevit etiam linguae Latinae corrumpentibus eam barbaris puritas.' Another author also saith that arts and learning did degenerate in these times from their genuine purity, together with the elegance of the Latin tongue, and that the empty babbling of Sophisters did everywhere make a noise in the schools." The authors referred to by Wood in this citation are, respectively, Leland and Bale.

[109] This is also supported by Thomas Wright, *A Volume of Vocabularies,* pp. xiii–xiv, privately printed, England, 1857.

[110] R. T. Gunther, *Early Science in Oxford,* Oxford, 1923, II, 63.

[111] Gunther, I, 336–337.

[112] Gunther, III, 152.

[113] P. Duhem, *Le Système du monde,* Paris, 1913–1916, IV, 157–164.

French technical expressions in geography and astronomy.[114]

In going through sources bearing on universities and education in general from 1347 to about 1375, it is plain that the Black Death and immediately succeeding outbreaks of pestilence had decided and dire consequences in these fields. These consequences appear in mortality among professors and other teachers, involving depreciation in number and calibre; in alarming decrease of students, and of ability on their part to provide themselves with an education; in difficulties and disappearance of universities themselves, owing to the desperate straits of cities and countries in which they had formerly flourished. Other forces were operative in producing these unhappy conditions, but, except for the Hundred Years' War in France, they seem to have been no more destructive or numerous than at earlier or later periods. Not only materially but morally was this a calamitous time for intellectual life in Europe. From students came severe arraignments of the inefficiency and unbridled cupidity of their masters, from church authorities charges of unbridled avarice and license of the clergy; while from many sources arose complaints of laxity and abuses among students and faculties. A deep mark, too, seems to have been left on the minds of men, changing the hopeful outlook of the thirteenth century to a spirit of gloom and anxiety, and a proneness to dwell upon disease and death.

Constructive efforts followed in the wake of the plague, and some beneficial changes. These were endowments of colleges, establishment of new universities and extension of prerogatives of the old, a growing recognition of student

[114] M. Cantor, *Vorlesungen über Geschichte der Mathematik*, Leipzig, 1913, II, 129.

rights, and a firmer foothold acquired by national languages in the fields of learning. But in spite of these it seems undeniable that the pestilence was definitely a blow to universities, education, and learning, and hence was to civilization a set-back of grave importance.

BIBLIOGRAPHY

The guiding principle in preparing this bibliography has been to make it as useful and convenient as possible. The arrangement is strictly alphabetical; the names of medieval authors, which appear with many variations in contemporary writings, have been Anglicized unless they are so well known in another form that to change them would be confusing; and, with similar exception, the medieval fashion of designating the individual by his first name has been retained. The great source collections, national publications of state documents and archives, biographical dictionaries, and bibliographies are not listed independently, except in the case of a few which were used systematically and exhaustively.

ALBERT, Consilium magistri Alberti ad pestilentiam in 1348. Edited by K. Sudhoff, in *Archiv für Geschichte der Medizin*, VI (1912), 316 f.

ALLBUTT, SIR T. C., The Historical Relations of Medicine and Surgery to the End of the Sixteenth Century; London, 1905.

——— Science and Medieval Thought; London, 1901.

ALLYN, H. B., The Black Death, Its Social and Economic Results, in *Annals* of *Medical History*, VII (1925), 226 ff.

ALFONSO OF CORDOVA, Epistola et regimen Alphontii Cordubiensis de pestilentia (1348?). Edited by K. Sudhoff, in *Archiv für Geschichte der Medizin*, III (1910), 224 ff.

ANDEL, M. A. VAN, Plague Regulations in the Netherlands, in *Janus*, XXI (1916), 410 ff.

ANSTEY, H., Munimenta Academica Oxoniensia; 2 Vols., London, 1868.

ARDERNE, JOHN, De arte physicali et de chirurgia. Translated by Sir D'Arcy W. Power; London, 1922.

——— Treatises of Fistula in Ano, Haemorrhoids, and Clysters; Translated by Sir D'Arcy W. Power; London, 1910.

AVICENNA (IBN SINA), Liber canonis de medicinis cordialibus et Cantica; Basel, 1556.

BALL, W. W. R., A Short Account of the History of Mathematics; London, 1888.

BALUZE, E., Miscellanea novo ordine digesta; Lucca, 1764.

——— Vitae paparum avenionensium, 1305–1394; edited by G. Mollat, 4 Vols., Paris, 1914–1922.

BÄR, M. (ED.), Urkunden und Akten zur Geschichte der Verfassung und Verwaltung der Stadt Koblenz bis zum Jahre 1500; Cologne, 1898.

BARBOT, J., Les Chroniques de la faculté de médecine de Toulouse du XIIIe au XXe siècle; 2 Vols., Toulouse, 1905.

BARNES, H., Visitations of the Plague in Cumberland and Westmoreland; Kendall, 1890.

BARTOLUS OF SASSOFERRATO, Commentaria; Venice, 1590–1602.

BASCOME, E., A History of Epidemic Pestilences; London, 1851.

BELLOC, HILAIRE, History of England; 3 Vols., London, 1925–1928.

BERTHELOT, P. E. M., La Chimie au moyen âge; 3 Vols., Paris, 1896.

BERTONI, G., I codici di un medico modenese del secolo XIV, in *Atti e memorie della Reale Deputatione di storia potria per le provincie modenesi*, Ser. V, Vol. IV (1905), pp. 125 ff.

BERTONI, G., AND E. P. VICINI, Medici modenese del secolo XIV, in *Atti e memorie della Reale Deputatione di storia patria per le provincie modenesi*, Ser. V, Vol. IV (1905), pp. 133 ff.

BINI, P. D. V., Memorie istoriche della Perugina Università degli studj e dei suoi professori; Perugia, 1816.

BONGI, S., Bandi lucchesi del secolo XIV; Bologna, 1863.

BOOS, H., Urkundenbuch der Stadt Worms; 4 Vols., Berlin, 1897–1901.

BOUDET, M., AND R. GRAND, Etude sur les épidémies de peste en Haute-Auvergne, XIVe-XVIIIe siècles; Paris, 1902.

BROCKELMANN, C., Geschichte der arabischen Litteratur; 2 Vols., Berlin, 1902.

BRODRICK, G. C., A History of the University of Oxford; 4th ed., London, 1900.

BUGIEL, V., La Médecine et les médecins dans les récits du nouvelliste espagnol Jean Manuel (1282–1348), in *Bulletin de la Societé Française d'histoire de la Médecine*, XVI, (1922), 293 ff.

CAIUS, JOHN, Annals of Gonville and Caius College; Cambridge, 1904.

——— De antiquitate Cantabrigiensis Academiae; London, 1568. New edition of his works, Cambridge, 1912.

Cajori, F., History of Mathematics; 2d ed., New York, 1919.

Calendar of the Patent Rolls, Vol. VIII (1348–1350); London, 1905.

Cambridge University, Documents Relating to the University and Colleges; 3 Vols., London, 1852.

Cantor, M., Vorlesungen über Geschichte der Mathematik; 4 Vols., 3d ed., Leipzig, 1899–1908. Reprint of Vol. II in 1913.

Carabellese, F., La peste del 1348 e le condizioni della santà pubblica in Toscana; Florence, 1897.

Carbonelli, G., Il "De sanitatis custodia" di Maestro Giacomo Albini da Moncaliere, con altri documenti sulla storia della medicina negli stati sabaudi nei secoli XIV e XV; Pinerolo, 1906.

Carlebach, E., Die rechtlichen und sozialen Verhältnisse der jüdischen Gemeinden: Speyer, Worms, und Mainz, von ihren Anfängen bis zur Mitte des 14. Jahrhunderts; Leipzig, 1901.

Casiri, M., Bibliotheca arabico-hispana escurialensis; 2 Vols. Madrid, 1770.

Cayla, P., L'Epidemie de peste en 1348 à Narbonne; Montpellier, 1906.

Cecchetti, B., Medicina in Venetia nel 1300, in *Archivio Veneto*, XXV (1883), 376 ff.

——— Vita del Veneziani nel secolo XIV, in *Archivio Veneto*, XXV (1883), 361 ff.; XXVI (1884), 77 ff., 251 ff.

Chavant, F., La Peste à Grenoble, 1410–1643; Lyons, Paris, 1903.

Chéreau, A., Les Médecins de six rois de France, 1270–1350, in *Union médicale*, N. S., XXIV (1864), 573 ff., 605 ff., 625 ff.

Chiapelli, A., Gli ordinamenti sanitari del Comune di Pistoia contro la peste del 1348, in *Archivio storico italiano*, Ser. IV, Vol. XX (1887), 6 ff.

Clark, A., The colleges of Oxford; London, 1891.

Cleu, H., Les maladies épidémiques et contagieuses en Lorraine du IXe au XIXe siècle, in *Bulletin de la Societé Française d'histoire de la médecine*, XIII (1914), 236 ff.

Colle, J., Medicina practica; 2 Vols. in 1, Pisa, 1617.

Collectanea, 3d Ser., London, 1896, M. Burrows and L. Smith, editors.

COOPER, C. H., Annals of Cambridge; Cambridge, 1842.

——— Memorials of Cambridge; 2 Vols., 2d ed., Cambridge, 1860.

COPPI, E., Le università italiane nel medio evo; 3d ed., Florence, 1886.

CORRADI, A., Annali delle epidemie in Italia dei primi tempi fino al 1850; Bologna, 1860.

COULTON, G. C., The Black Death; New York, 1930.

CRAWFORD, R., Plague and Pestilence in Literature and Art; Oxford, 1916.

CREIGHTON, C., History of Epidemics in Britain; 2 Vols., Cambridge, 1891.

DALLARI, U., I rotuli dei lettori legisti e artisti dello studio bolognese dal 1384–1789; 4 Vols., Bologna, 1888–1924.

DELAMBRE, J. B. J., Histoire de l'astronomie du moyen âge; Paris, 1819.

DENIFLE, H., Die Entstehung der Universitäten des Mittelalters bis 1400; Berlin, 1885.

——— La Guerre de cent ans et la désolation des églises, monastères, et hôpitaux en France; 2 Vols., Paris, 1897–1899.

——— Die Statuten der Juristen-Universität Bologna vom J. 1317–1347, und deren Verhältniss zu jenen Paduas, Perugias, Florenz, in *Archiv für Litteratur und Kirchengeschichte des Mittelalters*, III (1887), 196 ff.

——— Die Statuten der Juristen-Universität Padua vom Jahre 1331, in *Archiv für Litteratur und Kirchengeschichte des Mittelalters*, VI (1891–1892), 309 ff.

——— Les Universités françaises au moyen âge; Paris, 1872.

DENIFLE, H., AND A. CHATELAIN, Chartularium Universitatis Parisiensis; 4 Vols., Paris, 1889–1897.

DERENBOURG, HARTWIG, Les Manuscrits arabes de l'Escurial, Vol. 3 (E. Levi-Provençal, ed.), Paris, 1928.

Dictionary of National Biography.

DIONYSIUS SECUNDUS COLLE OF BELLUNA, De pestilentia 1348. 1350. et peripneumonia pestilentiali et maligna simul; published in J. Colle, *Medicina practica;* Pisa, 1617.

DÖLLINGER, J. I. J., JR., Betrachtungen über das Wesen der deutschen Universitäten; Wurzburg, 1920.

DORVEAUX, P., Régime contre la pestilence faict et composé par

Messieurs les médecins de la cité de Balle en Allemagne, in *Janus*, VI (1901).

Du Boulay, C. E., Historia Universitatis Parisiensis; 6 Vols., Paris, 1665–1673.

Duhem, P., Étude sur Léonard de Vinci; ceux qu'il a lus et ceux qui l'ont lu; 3 Vols., Paris, 1906–1913.

——— Le système du monde. Histoire des doctrines cosmologiques de Platon à Copernic; 5 Vols., Paris, 1913–1916.

Dupouy, E., Le Moyen âge médicale; Paris, 1895.

Dyer, G., History of the University and Colleges of Cambridge; London, 1814.

——— The Privileges of the University of Cambridge; London, 1824.

Ehrle, F., Die Ehrentitel der scholastischen Lehrer des Mittelalters, in *Bayerische Akademie der Wissenschaften, Sitzungsberichte,* IX (1919), 3 ff.

Eubel, C., Hierarchia catholica medii aevi; 3 Vols., 2d ed., Münster, 1901–1913.

Fabroni, A., Historia Academiae Pisanae; 3 Vols., Pisa, 1791–1795.

Facciolati, J., Fasti Gymnasii Patavini; Padua, 1757.

——— De Gymnasio Patavino syntagmata XII ex ejusdem gymnasii fastis excerpta; Padua, 1752.

Félibien, M., and G. A. Lobineau, Histoire de la ville de Paris; 5 Vols., Paris, 1725.

Felix Fabri, Historia Suevorum (*c.* 1480), in M. Goldast, *Suevicarum rerum scriptores aliquot veteres*; Frankfort, 1606.

Fitzralph, Richard, Defensorum curatorum contra eos qui privilegiatos se dicunt (1357), in E. Brown, *Fasciculus rerum expetendarum et fugiendarum*; London, 1690.

Ford, W. W., Quarantine, *Johns Hopkins Hospital Bulletin,* XXV (1914).

Fort, G. F., Medical Economy during the Middle Ages; New York, 1883.

Fournier, M., Les Statuts et privilèges des universités françaises depuis leur fondation jusqu'à 1789; 4 Vols., Paris, 1890–1894.

Frari, A. A., Della peste e della pubblica amministrazione sanitaria; Venice, 1840.

Friedell, E. Kulturgeschichte der Neuzeit. Die Krisis der

europäischen Seele von der schwarzen Pest bis zum Weltkrieg; 2 Vols., Münich, 1927.

Gabriel de Musis of Piacenza (d. 1356), Istoria de morbo sive mortalitate que fuit de 1348. Printed by Henschel in Haeser's *Archiv für gesammte Medizin*, II (1842), 25 ff., and by Tononi in *Giornale ligustico*, IX (1883), 139 ff.

Galen, C., Edited by K. G. Kühn, Medicorum graecorum opera quae extant; Vols. 1–20, Leipzig, 1821–1823.

Garrison, F. H., An Introduction to the History of Medicine. 4th ed., Philadelphia, 1929.

Gascoigne, Thomas, Liber veritatum (1433–1457), edited by J. E. T. Rogers, Loci e libro veritatum; Oxford, 1881.

Gasquet, F. A., The Great Pestilence; London, 1893, published as The Black Death; London, 1908.

Gentile of Foligno, Consilia contra pestilentiam, 1348. For manuscript and printed editions used, see notes in text, p. 9.

Gherardi, A., Statuti della università e studio fiorentino; Florence, 1881.

Gherardi S., Di alcuni materiali per la storia della facolta matematica nell' antica università di Bologna; Bologna, 1846.

Gloria, A., Monumenti della università di Padova, 1318–1405. 2 Vols., in *Padua Università Studi*, 1888.

——— Studi editi dalla Università di Padova, in *Istituto veneto memorie*, 22,[2] 22.[3]

Gross, C., The Political Influence of the University of Paris in the Middle Ages, in the *American Historical Review*, VI (1901), 440 ff.

Gunther, R. T., Early Science in Oxford; 7 Vols., Oxford, 1923–1930.

Guy of Chauliac, La Grande chirurgie de Gui de Chauliac; edited by E. Nicaise; Paris, 1890.

Haeser, H., Bibliotheca epidemiographica; 2d ed., Jena, 1862.

——— Lehrbuch der Geschichte der Medizin und der epidemischen Krankheiten; 3 Vols., 3d ed., Jena, 1876.

Haffner, P., ed., Frankfurter zeitgemässe Broschuren; Frankfort, 1880 ff.

Hanssen, P., Geschichte der Epidemien bei Menschen und Tieren im Norden; Glückstadt, 1925.

Hauréau, B., Histoire de la philosophie scholastique; 2 Vols. in 3, Paris, 1872–1880.

HECKER, J. F. C., Die grossen Volkskrankheiten des Mittelalters; edited by Hirsch. Berlin, 1865; a later edition of Der schwarze Tod im 14. Jahrhundert. Berlin, 1832. Translated by B. G. Babington, as Epidemics of the Middle Ages. 3d ed., London, 1859.

HEILBRONNER, J. C., Historia mathesos; Leipzig, 1742.

HEITZ, P., AND W. L., SCHREIBER, Pestblätter des XV. Jahrhunderts; Strasburg, 1901.

HENRY OF LAMMESPRINGE, Die magdeburger Schöppenchronik Heinrichs von Lammespringe; edited by K Janicke, in *Die Chroniken der niedersächsischen Städte,* Vol. VII, Leipzig, 1869.

HENRY OF MONDEVILLE, Die Chirurgie des Heinrichs von Mondeville; edited by J. Pagel; Berlin, 1890–1891. Translated into French by E. Nicaise, 1893.

HENSLOW, G., Medical Works of the Fourteenth Century; London, 1899.

HIRSCH, A., Biographisches Lexikon der hervorragenden Aerzte aller Zeiten und Völker; 6 Vols., Vienna, 1884–1888.

Histoire littéraire de la France; 36 Vols., Paris, 1733–1927.

HISTORICAL MANUSCRIPTS COMMISSION, Report on the Manuscripts Belonging to the Dean and Chapter of Canterbury, C. T. Martin, ed., in Fifth Report (1874), and Eighth Report (1881).

HOEHLBAUM, C., Mittheilungen aus dem Stadtarchiv von Köln; Cologne, 1882 ff.

———Regesten der Erzbischöfe von Mainz von 1289–1396; Cologne, 1907 ff.

HÖNIGER, R., Der schwarze Tod in Deutschland; Berlin, 1882.

———Gang und Verbreitung des schwarzen Todes; Berlin, 1881.

IBN AL-KHATĪB, A Very Useful Inquiry into the Horrible Sickness. Written about 1349, and published in the Arabic, with German translation, by M. J. Müller, in *Sitzungsberichte* der *Kgl. bayer Akad. d. Wiss. zu München, Philos.-Philol. Kl.* II (6 June, 1863), 1 ff.

IBN BATUTA, The Travels of Ibn Batuta; translated into English by S. Lee; London, 1829.

IBN KHĀTIMAH, Morbi in posterum vitandi descriptio et remedia. Written in 1349, and translated from the Arabic into German and published by Taha Dinânah, in *Archiv für Geschichte der Medizin*, XIX (1927), 30 ff.

Irsay, S. d', The Black Death and the Medieval Universities, in *Annals of Medical History*, VII (1925), 220 ff.

——— Notes to the origin of the expression "Atra mors," in *Isis*, VIII (1926), 328 ff.

James of Agramont, Epistola de Maestre Jacme d' Agramont als honorats e discrets seynors pahers e Conseyll de la ciutat de Leyda, 24 Abril, 1348. See text, p. 9.

Jessopp, A., The Black Death in East Anglia, in *The Coming of the Friars*. 8th ed., London, 1913.

John Hake, or Griese, of Göttingen. Prescriptions against the pestilence (1349?); edited by K Sudhoff, in *Archiv für Geschichte der Medizin*, V, 37 f.

John of Marignolli, or John of Florence, d. after 1357. Complete Latin text of travels in Fontes rerum bohemicarum, III, 492 ff.; partial English translation in Sir H. Yule, Cathay, new ed., III (1914), 177 ff.

John of Parma, Chronicle of the Pestilence from 1348 to 1377, published by A. Pezzana, in *Storia della città di Parma*, I, Parma, 1837.

John of Penna, Consilium magistri Johannis della Penna in magna pestilentia, 1348, published by K. Sudhoff, in *Archiv für Geschichte der Medizin*, V, 341 ff., and XVI, 162 ff.

John of Reading, Chronicon Johannis de Reading; edited by J. Tait, Manchester, 1914.

Kaufmann, G., Geschichte der deutschen Universitäten; Stuttgart, 1888.

Kink, R., Geschichte der Kaiserlichen Universität zu Wien; 2 Vols. Vienna, 1854.

Klebs, A. C., and E. Droz, Remèdes contre la peste; Paris, 1925.

Klebs, A. C., and K. Sudhoff, Die ersten gedruckten Pestschriften; Münich, 1926.

Knipping, R., ed., Die Regesten der Erzbischöfe von Köln im Mittelalter; Cologne, 1901 ff.

Krafft-Ebing, R. von, Zur Geschichte der Pest in Wien, 1349–1898; Leipzig, 1899.

Laval, V., Cartulaire de l'Université d'Avignon; Avignon, 1884.

LEACH, A. F., ED., Early Yorkshire Schools, Yorkshire Archaeological Society Records, Ser. 27, London, 1899.

LECHNER, K., Das grosse Sterben in Deutschland in den Jahren 1348 bis 1351, und die folgenden Pestepidemien bis zum Schlusse des 14. Jhts.; Innsbruck, 1884.

LEERSUM, E. C., VAN, Master Jan Yperman's "Cyrurgia," in *Janus*, XVIII (1913).

LEVETT, A. E., The Black Death on the Estates of the See of Winchester; Oxford, 1916.

LOWTH, R., Life of William of Wykeham; London, 1758.

LUDOLPH OF SAXONY, La grande vie de Jésus-Christ. Translation of M. P. Augustine; 4 Vols., Paris, 1864–1865. Later edition 1883, translated by Broquin.

MAIOCCHI, R., Codice diplomatico dell' Università di Pavia; Vol. I, Pavia, 1905.

MALAGOLA, C., Statuti delle università e dei collegi dello Studio Bolognese; Bologna, 1888.

MALLET, C. E., History of the University of Oxford. 3 Vols., London, 1924–1927.

MANACORDA, G., Storia della Scuola in Italia; il medio evo; 2 Vols., Milan, 1914.

MARTIN, K., Versuch einer geographischen Darstellung einiger Pestepidemien, in *Petermann's Mittheilungen aus Justus Perthes' geographischer Anstalt*, XIV (1879), 257 ff.

MARTINOTTI, G., L'insegnamento dell' anatomia in Bologna, pp. 3–146, Vol. II (1911), of *Studi e memorie per la storia dell' Università di Bologna*, Vols. 1–4, Bologna, Vol. 5, Modena, 1907–1920.

MAYCOCK, A. L., A Note on the Black Death, in *Nineteenth Century*, XCVII (Jan.–June, 1925), 456 ff.

MEINERS, C., Geschichte der Entstehung und der Entwickelung der hohen Schulen unsers Erdtheils; 4 Vols., Göttingen, 1802–1805.

——— Uber die Verfassung und Verwaltung deutscher Universitäten; Göttingen, 1801.

MEINSMA, K. O., De zwarte dood (1347–1352); Zutphen, 1924.

MICHAEL PLATIENSIS, Historia sicula ab anno 1337 ad annum 1361, in R. Gregorio, *Bibliotheca scriptorum qui res in Sicilia gestas retulere*, 1791–1792.

MICHON, L. A. J., Étude d'histoire médicale; documents inédits sur la grande peste de 1348; Paris, 1860.

MODE, P. G., Influence of the Black Death on the English Monasteries; Chicago, 1916.

MOHR, C. VON, Die Regesten der Archive der Schweizerischen Eidgenossenschaft; Berne, 1848.

MOLL, A., Der schwarze Tod in Würtemberg, in *Medicinisches Correspondenz-Blatt des würtembergischen ärtzlichen Vereins*, Vol. XXVII, Nos. 32, 33, 34, 40, 41, Stuttgart, 1857.

MONTPELLIER, PRACTITIONER, Tractatus de epidimia (1349), in L. A. J. Michon, *Documents inédits*. Paris, 1860.

MONTPELLIER, UNIVERSITY, Cartulaire de l'Université de Montpellier; 2 Vols., Montpellier, 1890–1912.

MOORE, N., History of St. Bartholomew's Hospital; 2 Vols., London, 1918.

——— History of the Study of Medicine in the British Isles; Oxford, 1908.

MORRIS, R., AND W. SKEAT, Specimens of Early English; Oxford, 1889.

MULLINGER, B., The University of Cambridge, from the Earliest Times to the Injunction of 1535; 2 Vols., Cambridge, 1873.

MURATORI, L. A., Rerum italicarum scriptores ab anno aerae Christianae 500 ad 1500; 25 Vols., Milan, 1723–1751. The references here are to the old Muratori, since the new edition which is now appearing is still incomplete.

NAPLES, UNIVERSITY, Storia della Università di Napoli; Naples, 1924.

NEUBURGER, M., Geschichte der Medizin; 2 Vols., Stuttgart, 1906–1911. Translation of Vol. I by E. Playfair, London, 1910, but with the omission of valuable notes and supplementary material.

NICAISE, E., L'Anatomie et la physiologie au XIVe siècle, in *Revue de chirurgie*, XIII (1893). See also Guy of Chauliac.

NOHL, J., Der Schwarze Tod, 1348–1720; Potsdam, 1924. Translated by C. H. Clarke, as The Black Death; London, 1926.

NORBERT, P. ED., Saint Jean Discalcéat, Frère Mineur, 1279–1349; an unedited manuscript of the fourteenth century, by an anonymous author; St.-Brieuc, 1911.

NORDI, L., AND E. ORIOLI, EDITORS. Chartularium studii bononiensis; 7 Vols., Bologna, 1909–1923.

OSLER, SIR W., Incunabula medica; London, 1923.

OZANAM, J. A. F., Histoire médicale générale et particulière des maladies épidémiques contagieuses et épizootiques, qui ont regné en Europe depuis les temps les plus réculés jusqu'à nos jours; 2d ed., Paris, 1835.

PAGEL, J., Geschichte der Heilkunde im Mittelalter, in *Handbuch der Geschichte der Medizin,* Vol. I, 622 ff., Jena, 1902.

PALMIERI, A., L'arte medica nell' antico Appenino bolognese, in *Atti e memorie della Regia deputatione di storia patria per le Romagne;* Ser. IV, Vol. I (1911), pp. 225 ff.

PANSIER, P., Les Maîtres de la faculté de médecine de Montpellier au moyen âge, in *Janus,* IX–X (1904–1905). Les Médecins des papes d'Avignon (1308–1403), in *Janus* XIV (1909), 405 ff.

PARIS, UNIVERSITY, Compendium de epidimia per Collegium Facultatis Medicorum Parisius, 1348. For MSS and editions, see text, pp. 14–15.

PAULSEN, F., Die Gründung der deutschen Universitäten im Mittelalter, in *Historische Zeitschrift,* XLV (1881), 251 ff.

——— Organisation und Lebensordnungen der deutschen Universitäten im Mittelalter, in *Historische Zeitschrift,* XLV (1881), 385 ff.

PETERS, H., Der Arzt und die Heilkunst in der deutschen Vergangenheit; Leipzig, 1900.

PETRARCH, F., Opera quae extant omnia; Basel, 1581.

PHILIPPE, A., Histoire de la Peste Noire; Paris, 1853.

POWER, SIR D'ARCY W. The Lesser Writings of John Arderne; London, 1913. See also Arderne, John.

——— Memorials of the Craft of Surgery in England; London, 1886.

——— A Short History of St. Bartholomew's Hospital; London, 1923.

PRANTL, K. VON, Geschichte der Logik im Abendlande; 4 Vols., Leipzig, 1855–1870.

PREZZINER, G., Storia del pubblico studio e delle Società scientifiche e letterarie di Firenze; 2 Vols., Florence, 1810.

PUSCHMANN, T., History of Medical Education; translated by E. H. Hare; London, 1891.

PUTNAM, B., Enforcement of the Statute of Laborers during the First Decade after the Black Death; New York, 1908.

——— Wage Laws for Priests, in *American Historical Review*, XXI (1915), 12 ff.

RASHDALL, H., Universities of Europe in the Middle Ages; Oxford, 1895.

RÉBOUIS, H. E., Étude historique et critique sur la peste; Paris, 1888.

REES, W., The Black Death in Wales, in *Transactions of the Royal Historical Society*, Ser. IV, Vol. III (1920), 115 ff.

REH, H., Zur mittelalterlichen Kulturgeschichte, in *Bücherei der Volkshochschule*, Vol. 1, No. 5, Würzburg, 1920.

RICHER, P., L'Art et la médecine; Paris, 1902.

ROTHSCHILD, L., Die Judengemeinden zu Mainz, Speyer, und Worms von 1349–1438; Berlin, 1904.

SAHM, W., Geschichte der Pest in Ostpreussen; Leipzig, 1905.

SALTER, H. E., Medieval Archives of the University of Oxford; 2 Vols., Oxford, 1920.

SANDONNINI, T., Di un codice del XIV secolo e dell' antico studio modenese, in *Atti e memorie della Reale deputatione di storia patria per le provincie modenesi*, Ser. V, Vol. IV (1905), 81 ff.

SARTI, M., AND M. FATTORINI, De claris archigymnasii bononiensis professoribus a seculo XI usque ad seculum XIV; 3 Vols., Bologna, 1888–1896.

SARTON, G., Introduction to the History of Science; Vol. I, Baltimore, 1927.

SAUERLAND, H. V., ED., Urkunden und Regesten zur Geschichte der Rheinlande aus dem vatikanischen Archiv; Cologne, 1902–1913.

SAVIGNY, F. K. VON, Geschichte des römischen Rechts im Mittelalter; 7 Vols., 2d ed., Heidelberg, 1834–1851.

SAVONAROLA, M., Commentariolus de laudibus patavinis, written about 1440, published in Muratori, *Scriptores*, XXIV (1728).

SCHLIKENRIEDER, J. J., Chronologia diplomatica Universitatis Viennensis; Vienna, 1753.

SCHMIDT, G., ED., Päpstliche Urkunden und Regesten aus den Jahren 1295–1352, die Gebiete der heutigen Provinz Sachsen und deren Umlande betreffend; Halle, 1886.

SCHMITZ, H. J., Das Volksschulwesen im Mittelalter, in Haffner's *Frankfurter zeitgemässe Bröschuren*, Vol. II, Frankfort, 1881.

SCHOPPLER, H., Über regensburger Aerzte, in *Archiv für Geschichte der Medizin,* XV (1923), 105 ff.

SCHWESTERMÜLLER, K., Regiment und Lehre wider die schwere Krankheit der Pestilenz; Berlin, 1925.

SEEBOHM, F., The Black Death and Its Place in English History, in the *Fortnightly Review,* II (1865), 149 ff.

SEIDEL, E., Die Lehre der Kontagion bei den Arabern, in *Archiv für Geschichte der Medizin,* VI (1912), 81 ff.

SIMON OF COVINO, De judicio solis in convivio Saturni, written in 1350, and published by Littré E., in *Bibliothéque de l'Ecole de Chartes,* Ser. I, Vol. II (1840–1841), pp. 206 ff.

SIMON OF PHARES, "Recueil des plus célèbres astrologues et quelques hommes doctes" faict par Symon de Phares, du temps de Charles VIII'; edited by E. Wickersheimer; Paris, 1929.

SIMONINI, R., Maino de Maineri ed il suo "Libellus de praeservatione ab epydimia;" Modena, 1923.

SINGER, D. W., Some Plague Tractates (Fourteenth and Fifteenth Centuries); London, 1916.

SISAM, K., Fourteenth Century Verse and Prose; Oxford, 1923.

SMET, J.-J. DE, ED., Breve chronicon Flandriae clerici anonymi (1333–1356), in *Recueil des chroniques de Flandre,* III, 14 ff., Brussels, 1837–1865.

STEIN, W., ED., Akten zur Geschichte der Verfassung und Verwaltung der Stadt Köln im 14. und 15. Jahrhundert; Cologne, 1893.

STICKER, G., Abhandlungen aus der Seuchengeschichte und Seuchenlehre; 2 Vols. in 3, Giessen, 1908–1912.

SUDHOFF, K., Pestschriften aus den ersten 150 Jahren nach der Epidemie des Schwarzen Todes, in *Archiv für Geschichte der Medizin,* Vols. II–XVII (1909–1925).

THORNDIKE, L., The Blight of Pestilence on Early Modern Civilization, in *American Historical Review,* XXXII (1927), No. 3, 455 ff.

——— History of Magic and Experimental Science; 2 Vols., New York, 1923.

——— Science and Thought in the Fifteenth Century; New York, 1929.

——— Some Vatican manuscripts of Pest Tractates, in *Archiv für Geschichte der Medizin,* Vol. XXII (1929), No. 2, pp. 200 f.

Tiraboschi, G., Biblioteca modenese; 6 Vols., Modena, 1781–1786.

——— Storia della letteratura italiana; 16 Vols., Milan, 1822–1826.

Vergerius, P. P., Vitae principum carrariensium, published by Muratori, *Scriptores*, Vol. XVI.

Viard, J., Journaux du trésor de Philippe VI de Valois, in *Documents inédits de l'histoire de France*, 1899.

Villani, Giovanni, Cronica universale, with continuation after 1348 by Matteo Villani, in Muratori, *Scriptores*, XIII–XIV.

Villani, P., Le vite d'uomini illustri fiorentini, published in Vol. VI of the *Cronica di Matteo Villani*, Florence, 1826.

Volta, Z., Dei gradi academici conferiti nello "Studio Generale" di Pavia sotto il dominio Visconteo, in *Archivio storico lombardo*, Ser. II (1890).

Vossius, G. J., De universae matheseos natura et constitutione liber; Amsterdam, 1650.

Weber, F. P., Aspects of Death in Art; 2d ed., London, 1914.

Wenck, K., Johann von Göttingen (d. 1349), in *Archiv für Geschichte der Medizin*, XVII (1925), 141 ff.

Wickersheimer, E., L'Anatomie de Guido de Vigevano, in *Archiv für Geschichte der Medizin*, VII (1913), 1 ff.

——— Les Maladies épidémiques ou contagieuses (peste, lèpre, syphilis) et la Faculté de Médecin de Paris, de 1399 a 1511, in *Bulletin de la Societé française d'histoire de la médecine*, XIII (1914), 21 ff.

——— Les médecins de la nation anglaise (ou allemande) de l'Université de Paris au XIVe et XVe siècles, in *Bulletin de la* Societé française d'histoire médecine, XII (1913), 285 ff., 537 ff.

——— La Peste Noire à Strasbourg et le régime des cinq médecins strasbourgeois, in *Proceedings of the Third International Congress of the History of Medicine*, p. 54 ff., Antwerp, 1923.

Wilkins, D., Concilia Magnae Britanniae et Hiberniae; 4 Vols., London, 1737.

Wilson, F. P., ed., Plague Pamphlets of Thomas Dekker; Oxford, 1925.

Wood, A., The History and Antiquities of the University of Oxford; 2 Vols., Oxford, 1792.

Woolf, C. N. S., Bartolus, His Position in the History of Medieval Thought; Cambridge, 1918.

Workman, H. B., John Wyclif; 2 Vols., Oxford, 1926.
Wyclif, John, De ecclesia; edited by J. Loserth; London, 1886.
Young, S., The Annals of the Barber-Surgeons of London; London, 1890.

INDEX

COLUMBIA UNIVERSITY PRESS

Columbia University

New York

FOREIGN AGENTS

OXFORD UNIVERSITY PRESS

Humphrey Milford

Amen House, London, E. C.

Bei Fragen zur Produktsicherheit wenden Sie sich bitte an:
If you have any questions regarding product safety,
please contact:

Walter de Gruyter GmbH
Genthiner Straße 13
10785 Berlin
productsafety@degruyterbrill.com